The Scalar–Tensor Theory of Gravitation

The scalar–tensor theory of gravitation is one of the most popular alternatives to Einstein's theory of gravitation. This book provides a clear and concise introduction to the theoretical ideas and developments, exploring scalar fields and placing them in context with a discussion of Brans–Dicke theory. Topics covered include the cosmological constant problem, time-variability of coupling constants, higher-dimensional space-time, branes, and conformal transformations. The authors emphasize the physical applications of the scalar–tensor theory and thus provide a pedagogical overview of the subject, keeping more mathematically detailed sections for the appendices.

This book is suitable as a textbook for graduate courses in cosmology, gravitation, and relativity. It will also provide a valuable reference for researchers.

YASUNORI FUJII received his PhD, on the analogy between the strong interaction and the electromagnetic interaction, from Nagoya University in 1959. Between 1963 and 1992 he did research on the theory of particle physics and gravity, including pioneering work on the idea of non-Newtonian gravity, at the Institute of Physics, University of Tokyo. During this period, he also spent two years at Stanford University and a year at Purdue University. He is currently emeritus professor at the University of Tokyo and continues to pursue his research interests at the Nihon Fukushi University. Professor Fujii is the author of several books in Japanese, both for expert physicists and for general readers. He has also published numerous articles and papers in leading journals.

KEI-ICHI MAEDA received his PhD from Kyoto University in 1980. He and his contemporaries created a new research group in Kyoto, which was at the root of numerical relativity research in Japan. In 1983 he became a postdoctoral student at SISSA, Trieste, working under Dennis Sciama. He moved to the Meudon Observatory in Paris in 1987 and worked on black hole solutions in string theory. In 1989 Professor Maeda became affiliated with the Department of Physics at Waseda University. Since 1998, he has been the associate editor of the *Journal of General Relativity and Gravitation*, and also the vice-chief editor of the *Journal of the Physical Society of Japan* since 2001. In addition to many important journal articles, Professor Maeda has written several popular science books in Japanese.

T0185807

CAMBRIDGE MONOGRAPHS ON
MATHEMATICAL PHYSICS

General editors: P. V. Landshoff, D. R. Nelson, S. Weinberg

[†] Issued as a paperback

The Scalar–Tensor Theory
of Gravitation

YASUNORI FUJII

Nihon Fukushi University

KEI-ICHI MAEDA

Waseda University, Tokyo

CAMBRIDGE
UNIVERSITY PRESS

CAMBRIDGE UNIVERSITY PRESS
Cambridge, New York, Melbourne, Madrid, Cape Town, Singapore, São Paulo

Cambridge University Press
The Edinburgh Building, Cambridge CB2 8RU, UK

Published in the United States of America by Cambridge University Press, New York

www.cambridge.org
Information on this title: www.cambridge.org/9780521811590

First published 2003
This digitally printed version 2007

A catalogue record for this publication is available from the British Library

Library of Congress Cataloguing in Publication data
Fujii, Yasunori.
The scalar–tensor theory of gravitation / Yasunori Fujii, Kei-ichi Maeda.
p. cm. – (Cambridge monographs on mathematical physics.)
Includes bibliographical references and index.
ISBN 0 521 81159 7
1. Gravitation. I. Maeda, Kei-ichi, 1950– II. Title. III. Series.
QC178 .F85 2003
531'.14–dc21 2002067367

ISBN 978-0-521-81159-0 hardback
ISBN 978-0-521-03752-5 paperback

To

Kaoru and Reiko

Contents

Preface

During the last few decades of the twentieth century, we saw an almost triumphant success in establishing that Einstein's general relativity is correct, both experimentally and theoretically. We find nevertheless considerable efforts still being made in terms of "alternative theories." This trend may be justified insofar as the scalar–tensor theory is concerned, as will be argued, not to mention one's hidden desire to see nature's simplest imaginable phenomenon, a scalar field, be a major player.

The success on the theoretical front prompted researchers to study theories with the aim of unifying gravitation and microscopic physics. Among them string theory appears to be the most promising. According to this theory, the graviton corresponding to the metric tensor has a scalar companion, called the dilaton. The interaction between these two fields is surprisingly similar to what Jordan foresaw nearly half a century ago, without sharing ideas that characterize the contemporary unification program. There seems to be, however, a crucial point that might constrain the original proposal through the value of the parameter ω, whose inverse measures the strength of the coupling of the scalar field.

More specifically, string theory predicts that $\omega = -1$, which goes against the widely accepted constraint from observation, namely $\omega \gtrsim 10^3 \gg 1$. Although many more details have yet to be worked out in order for string theory to be compared with the real world, we point out that expecting the dilaton to be close to the limit of total decoupling is by no means obvious or natural. One may even suspect that the scalar force has a finite force-range, rendering the solar-system experiments from which a large value of ω derives irrelevant.

Also related is the fact that string theory, like other unification theories, allows coupling to matter at the level of the Lagrangian, thus inevitably violating the weak equivalence principle (WEP), in contrast to what Brans and Dicke proposed as a modification of Jordan's model.

Both of these considerations, together with our own study of the cosmological equations, suggest a possible way out by revisiting the idea of non-Newtonian gravity, which might be present somewhat below the constraints obtained so far, as a manifestation of the scalar field.

There are other aspects of the scalar–tensor theory requiring more careful understanding, including such issues as how the physical "conformal frame" is singled out, and how much time-variability of the gravitational constant there could be. There is also the question arising from the sign of the energy of the scalar field in the original conformal frame, placing string theory further away from the near-complete decoupling.

One of the purposes of this book is to provide detailed accounts of subtleties that might have escaped attention, which are based on naive questions but need to be treated with sufficient caution. We find that this theory, with appropriate modifications, seems to provide a small window through which we can look into what the expected unification theory is.

We then ask whether there are any observational signals that make such a departure from the standard theory urgent. We may consider a modern version of the problem of the cosmological constant as well as observational searches for time-variability of the coupling constants. From this point of view, we apply the scalar–tensor theory to a cosmology with Λ in accordance with the recent discovery from type-Ia supernovae. There has been an expectation, which is sound but nonetheless somewhat vague, that a light scalar field with a universal coupling should play a role. We offer what we hope is a realistic implementation.

We discuss another more phenomenological approach, which is based on a scalar field now widely known as "quintessence." Also included is a brief introduction to "brane cosmology," which might provide a new breed of cosmological model descending from higher-dimensional space-time.

Toward the end of the book we attempt to relate the problem of the cosmological constant to the possible time-dependence of the fine-structure constant. Even though it is still provisional, this argument illustrates how the scalar–tensor theory, a highly constrained theory, has the capability of linking two otherwise disparate phenomena.

Since we focus on a limited range of subjects, we have not attempted to make the book encyclopedic.

We have dealt with some of the technical complications in 14 appendices so that readers can gain a good overview of the underlying flow of our story without being deterred by these complexities. They belong basically to the chapters as shown:

Chapter 1:   A, B

Chapter 2:   C, D, E, F

Chapter 3:   G, H

Chapter 4: J, K

Chapter 5: L

Chapter 6: M, N, O.

Some of them might be skipped entirely. We hope, on the other hand, that some could serve as problems or exercises. Appendix D, for example, might be an answer to the following problem: "Show that the matter energy–momentum tensor is covariantly conserved if $\mathcal{L}_{\text{matter}}$ is independent of ϕ." Appendix L was prepared particularly for readers unfamiliar with "brane" geometry. Appendix C is truly fundamental throughout the book. Appendix A might be used as a basis of part of section 5.3, as might Appendix L.

The book is a consequence of discussions with and critical comments from many of our friends and colleagues. Among them we wish to express our thanks particularly to Yuichi Chikashige, Yon-Min Cho, Ephraim Fischbach, Shoichi Ichinose, Takashi Ikegami, Satoru Ikeuchi, Susumu Kamefuchi, Mitsuhiro Kato, Shinsaku Kitakado, Kazuaki Kuroda, Shuntaro Mizuno, Masahiro Morikawa, Wei-Tou Ni, Janis Niedra, Tsuyoshi Nishioka, Nobuyoshi Ohta, Minoru Omote, Takeshi Saito, Misao Sasaki, Tetsuya Shiromizu, Naoshi Sugiyama, Akira Tomimatsu, and Tamiaki Yoneya for their invaluable help in shaping our basic attitude toward the subject.

Our thanks are also due to Humitaka Sato for his having suggested that one of us (Y. F.) begin writing this book. Our gratitude goes also to Yasushi Takahashi who compiled a series including a book in Japanese, whose title translates as *Gravitation and Scalar Field*, by Y. F., which was published by Kodan-sha in 1997, an outgrowth of which constituted a major part of the present book.

For the writing of this book, we consulted Jordan's book to ascertain how it differed from the ensuing paper by Brans and Dicke. A consequence was our adding a short passage in Chapter 1. We leave more stories to a forthcoming publication by Carl Brans, to whom we express our gratitude for his comments and having shown us his thesis.

We thank Owen Parkes, and Bonnie and Patrick Ion for helping us by correcting our English during the earlier periods of our work.

Y. Fujii
K. Maeda

Yokohama
March 2002

Conventions and notation

Greek indices run from 0 through 3, where $x^0 = ct$. The Minkowskian metric is diagonal $(-1, +1, +1, +1)$. Indices with overbars are those for D dimensions. Some of the quantities in higher dimensions are attached as a superscript (D) to the left, but not strictly all the time, as long as no confusion is expected to ensue.

We use the reduced Planckian system of units by choosing $c = \hbar = M_{\mathrm{P}} \, (= [8\pi G/(c\hbar)]^{-1/2}) = 1$, yielding the units of length, time, and energy given, respectively, by

$$8.07 \times 10^{-33}\,\mathrm{cm}, \quad 2.71 \times 10^{-43}\,\mathrm{s}, \quad 2.44 \times 10^{18}\,\mathrm{GeV}.$$

In this unit of time, the present age of the universe $(1.1\text{--}1.4) \times 10^{10}$ years is expressed as $10^{60.11}\text{--}10^{60.21}$.

The Christoffel symbol is defined as usual:

$$\Gamma^\lambda{}_{\mu\nu} = \tfrac{1}{2} g^{\lambda\rho}(\partial_\mu g_{\rho\nu} + \partial_\nu g_{\rho\mu} - \partial_\rho g_{\mu\nu}).$$

The same definition applies to higher dimensions as well, but simply with overbars in the indices.

The Riemann curvature tensor is defined by

$$R^\rho{}_{\sigma,\mu\nu} = \partial_\mu \Gamma^\rho{}_{\sigma\nu} + \Gamma^\rho{}_{\lambda\mu}\Gamma^\lambda{}_{\sigma\nu} - (\text{terms with } \mu \leftrightarrow \nu).$$

The Ricci tensor and scalar curvature are derived as

$$R_{\mu\nu} = R^\rho{}_{\mu,\rho\nu} \quad \text{and} \quad R = R^\mu{}_\mu,$$

respectively. $R > 0$ for a sphere.

Symbols used multiply

We followed the usual usage of symbols as much as we could. We could not avoid, however, using some of them for two or more different purposes. Some of them might be worth listing.

σ Originally the fluctuating part of the scalar field ϕ in the scalar–tensor theory (Chapter 2), but also the scalar field in the E frame, and further used to denote a scalar field in the quintessence models.

Φ The dilaton field in string theory (Chapter 1), but used also for the scalar field as a simplified representative of (nongravitational) matter fields (Chapters 4–6).

χ Decomposition of the metric (Chapter 2), also the second scalar field in the two-scalar model (Chapter 5).

q The exponent of the power of the scalar field ϕ which multiplies Λ in Chapter 4. The same symbol is used also as a negative exponent of σ in the inverse-power potential of the quintessence field in Chapter 5.

Other special symbols

ξ Related to the original notation ω by $\xi = |\omega^{-1}|/4$.

ϵ The sign of ω, agreeing also with the sign of the kinetic energy of the scalar field ϕ in the J frame.

A The radius of internal space in Appendix A for the Kaluza–Klein theory, also with $\mathcal{A} = A^2$. Similar usage is found in Appendix B as well. On the other hand, the scale factor of the universe is denoted by a.

D The dimensionality of space-time. We also use $d = D/2$.

1
Introduction

We begin this chapter with an overview in section 1 of how the scalar–tensor theory was conceived, how it has evolved, and also what issues we are going to discuss from the point of view of such cosmological subjects as the cosmological constant and time-variability of coupling constants. In section 2 we provide a simplified view of fundamental theories which are supposed to lie behind the scalar–tensor theory. Section 3 includes comments expected to be useful for a better understanding of the whole subject. This section will also summarize briefly what we have achieved.

In section 1 we emphasize that the scalar field in what is qualified to be called the scalar–tensor theory is not simply added to the tensor gravitational field, but comes into play through the nonminimal coupling term, which was invented by P. Jordan. Subsequently, however, a version that we call the prototype Brans–Dicke (BD) model has played the most influential role up to the present time. We also explain the notation and the system of units to be used in this book.

The list of the fundamental ideas sketched in section 2 includes the Kaluza–Klein (KK) theory, string theory, brane theory as the latest outgrowth of string theory, and a conjecture on two-sheeted space-time. Particular emphasis is placed on showing how closely the scalar field can be related to the "dilaton" as a partner of the graviton in string theory that emerged from an entirely different point of view.

Section 3 will be a collection of comments. We wish to answer potential questions that might be asked by readers who have not yet entered the main body of the book but have nonetheless heard something about the scalar–tensor theory. We also embed certain abstracts or advertisements of the related subjects which will be discussed later in detail. As a result of our doing so we expect that readers may be acquainted beforehand with

1

our achievements from a wider perspective of the whole development. The topics will cover the weak equivalence principle (WEP), parameters of the prototype BD model, conformal transformation, Mach's principle and variable G, and a question about whether there is any advantage to be gained from sticking to a complicated scalar–tensor theory instead of less constrained theories of scalar fields, like the quintessence model.

1.1 What is the scalar–tensor theory of gravitation?

Einstein's general theory of relativity is a geometrical theory of space-time. The fundamental building block is a metric tensor field. For this reason the theory may be called a "tensor theory." A "scalar theory" of gravity had earlier been attempted by G. Nodström by promoting the Newtonian potential function to a Lorentz scalar. Owing to the lack of a geometrical nature, however, the equivalence principle (EP), one of the two pillars supporting the entire structure of general relativity, was left outside the aim of the theory in the early 1910s. This did not satisfy Einstein, who eventually arrived at a dynamical theory of space-time geometry. His theory must have appeared highly speculative at first, but proved later to be truly realistic, since it was supported by observations of diverse physical phenomena, including those in modern cosmology. It also served as an excellent textbook showing how a new way of thinking develops to reality.

In spite of the widely recognized success of general relativity, now called the standard theory of gravitation, the theory has also nurtured many "alternative theories" for one reason or another. Among them we focus particularly on the "scalar–tensor theory." It might appear as if the old idea of scalar gravity were being resurrected. In fact, however, this type of theory does not merely combine the two kinds of fields. It is built on the solid foundation of general relativity, and the scalar field comes into play in a highly nontrivial manner, specifically through a "nonminimal coupling term," as will be explained shortly.

The scalar–tensor theory was conceived originally by Jordan, who started to embed a four-dimensional curved manifold in five-dimensional flat space-time [1]. He showed that a constraint in formulating projective geometry can be a four-dimensional scalar field, which enables one to describe a space-time-dependent gravitational "constant," in accordance with P. A. M. Dirac's argument that the gravitational constant should be time-dependent [2], which is obviously beyond what can be understood within the scope of the standard theory. He also discussed the possible connection of his theory with another five-dimensional theory, which had been offered by Th. Kaluza and O. Klein [3]. On the basis of these considerations he presented a general Lagrangian for the scalar field living in four-dimensional curved space-time:

$$\mathcal{L}_{\mathrm{J}} = \sqrt{-g}\left[\varphi_{\mathrm{J}}^{\gamma}\left(R - \omega_{\mathrm{J}}\frac{1}{\varphi_{\mathrm{J}}^{2}}g^{\mu\nu}\,\partial_{\mu}\varphi_{\mathrm{J}}\,\partial_{\nu}\varphi_{\mathrm{J}}\right) + L_{\mathrm{matter}}(\varphi_{\mathrm{J}}, \Psi)\right], \quad (1.1)$$

where $\varphi_{\mathrm{J}}(x)$ is Jordan's scalar field, while γ and ω_{J} are constants, also with Ψ representing matter fields collectively. The introduction of the *nonminimal coupling term*, $\varphi_{\mathrm{J}}^{\gamma}R$, the first term on the right-hand side, marked the birth of the scalar–tensor theory. The term $L_{\mathrm{matter}}(\varphi_{\mathrm{J}}, \Psi)$ was for the matter Lagrangian, which depends generally on the scalar field, as well.

For later convenience, we here explain the unit system we are going to use throughout the book. Since we always encounter relativity and quantum mechanics in the area of particle physics, it is convenient to choose a unit system in which c and $\hbar$ are set equal to unity. By doing so we can express all three fundamental dimensions only in terms of one remaining dimension, which may be chosen as length, time, mass, or energy. In particular, the gravitational constant, or Newton's constant, G turns out to have a mass dimension -2, or length squared. We then write

$$8\pi G = c\hbar M_{\mathrm{P}}^{-2}, \quad (1.2)$$

with M_{P} called the *Planck mass*, which is estimated to be 2.44×10^{18} GeV, which is quite heavy compared with other ordinary particles. We hereafter choose $M_{\mathrm{P}} = 1$. In this way we can express every quantity as if it were dimensionless. This unit system is called the *reduced Planckian unit system*, though $G = 1$ is often chosen in the *plain* Planckian unit system. We prefer the former system with the difference of $\sqrt{8\pi}$. We show units of length, time, and energy in this system expressed in conventional units:

$$8.07 \times 10^{-33}\,\mathrm{cm}, \quad 2.71 \times 10^{-43}\,\mathrm{s}, \quad 2.44 \times 10^{18}\,\mathrm{GeV}. \quad (1.3)$$

Sometimes, however, it is convenient to leave one of the dimensions still "floating," not necessarily set fixed. We choose it to be mass, for example, as was shown in (1.2). In the same context, the Lagrangian is found to have a mass dimension 4, while a derivative contributes a mass dimension 1. The metric tensor is dimensionless. If a scalar field has a conventional canonical kinetic term, then we conclude that the field has a mass dimension 1.

Now the second term on the right-hand side of (1.1) resembles a kinetic term of φ_{J}. Requiring this term to have a correct mass dimension 4 yields the result that φ_{J} has mass dimension $2/\gamma$. It then follows that $\varphi_{\mathrm{J}}^{\gamma}$, which multiplies R in the first term on the right-hand side of (1.1), has mass dimension 2, the same as G^{-1}. In this way we re-assure ourselves that the first two terms on the right-hand side of (1.1) contain no dimensional constant. This remains true for any value of γ, although this "invariance" under a change of γ need not be respected if $\varphi_{\rm J}$ enters the matter Lagrangian, in general.

Jordan's effort was taken over particularly by C. Brans and R. H. Dicke. They defined their scalar field φ by

$$\varphi = \varphi_{\rm J}^{\gamma}, \tag{1.4}$$

which simplifies (1.1) by making use of the fact that the specific choice of the value of γ is irrelevant, as explained above. This process is justified only because they demanded that the matter part of the Lagrangian $\sqrt{-g}L_{\rm matter}$ be decoupled from $\varphi(x)$ as an implementation of their requirement that the WEP be respected, in contrast to Jordan's model. The reason for this crucial decision, after the critical argument by Fierz [4] and others, will be made clear soon.

In this way they proposed the basic Lagrangian

$$\mathcal{L}_{\rm BD} = \sqrt{-g}\left(\varphi R - \omega\frac{1}{\varphi}g^{\mu\nu}\,\partial_\mu\varphi\,\partial_\nu\varphi + L_{\rm matter}(\Psi)\right). \tag{1.5}$$

We call the model described by (1.5) the *prototype BD model* throughout this book [5]. The adjective "prototype" emphasizes the unique features that characterize the original model compared with many extended versions. The dimensionless constant ω is the only parameter of the theory.

Note that we left out the factor 16π that multiplied the whole expression on the right-hand side in the original paper, to make the result appear in a more standard fashion. For this reason our φ is related to their original scalar field, denoted here by $\phi_{\rm BD}$, by the relation

$$\varphi = 16\pi\phi_{\rm BD}. \tag{1.6}$$

We now take a special look at the nonminimal coupling term, the first term on the right-hand side of (1.5). This replaces the Einstein–Hilbert term,

$$\mathcal{L}_{\rm EH} = \sqrt{-g}\frac{1}{16\pi G}R, \tag{1.7}$$

in the standard theory, in which R is multiplied by a constant G^{-1}. By comparing (1.5) and (1.7) we find that this model has no gravitational "constant," but is characterized by an *effective gravitational constant* $G_{\rm eff}$ defined by

$$\frac{1}{16\pi G_{\rm eff}} = \varphi, \tag{1.8}$$

as long as the dynamical field φ varies sufficiently slowly. In particular we may expect that φ is spatially uniform, but varies slowly with cosmic time, as suggested by Dirac. We should be careful, however, to distinguish G_{eff}, the gravitational constant for the tensor force only, from the one including the possible contribution from the scalar field, as will be discussed later.

As another point, we note that the second term on the right-hand side of (1.5) appears to be a kinetic term of the scalar field φ, but looks slightly different. First, the presence of φ^{-1} seems to indicate a singularity. Secondly, there is a multiplying coefficient ω. These are, however, superficial differences. The whole term can be cast into the standard *canonical* form by redefining the scalar field.

For this purpose we introduce a new field ϕ and a new dimensionless constant ξ, chosen to be positive, by putting

$$\varphi = \tfrac{1}{2}\xi\phi^2 \tag{1.9}$$

and

$$\epsilon\xi^{-1} = 4\omega, \tag{1.10}$$

in terms of which the second term on the right-hand side of (1.5) is re-expressed in the desired form;

$$\sqrt{-g}\left(-\tfrac{1}{2}\epsilon g^{\mu\nu}\,\partial_\mu\phi\,\partial_\nu\phi\right), \tag{1.11}$$

with $\epsilon = \pm 1 = \text{Sign}\,\omega$.

No singularity appears, as suggested. $\epsilon = +1$ corresponds to a normal field having a positive energy, in other words, not a *ghost*. Note that (1.11) becomes $\dot{\phi}^2/2$ for $\epsilon = +1$ in the limit of flat space-time where $g^{00} \sim \eta^{00} = -1$. The choice $\epsilon = -1$ seems to indicate a negative energy, which is unacceptable physically. As will be shown later in detail, however, this need not be an immediate difficulty owing to the presence of the nonminimal coupling. We will show in fact that some of the models do require $\epsilon = -1$. Even the extreme choice $\epsilon = 0$, corresponding to choosing $\omega = 0$ in the original formulation, according to (1.10), leaving ξ arbitrary, may not be excluded immediately. Note also that (1.9) shows that ϕ has a mass dimension 1, as in the usual formulation.

In this way (1.5) is cast into the new form

$$\mathcal{L}_{\text{BD}} = \sqrt{-g}\left(\tfrac{1}{2}\xi\phi^2 R - \tfrac{1}{2}\epsilon g^{\mu\nu}\,\partial_\mu\phi\,\partial_\nu\phi + L_{\text{matter}}\right). \tag{1.12}$$

Discussing consequences of (1.12) will be the main purpose of Chapter 2. We briefly outline here subjects of particular interest.

Obviously (1.12) describes something beyond what one would obtain simply by adding the kinetic term of the scalar field to the Lagrangian with the Einstein–Hilbert term. We reserve the term "scalar–tensor theory" specifically for a class of theories featuring a nonminimal coupling term or its certain extension.

As we explain in the subsequent section, there are theoretical models to be categorized in this class. More general models have also been discussed. The prototype BD model still deserves detailed study, from which we may learn many lessons useful in analyzing other models.

Deriving field equations from (1.12) is a somewhat nontrivial task, as we elaborate in Chapter 2. In particular, we obtain

$$\Box\varphi = \zeta^2 T, \tag{1.13}$$

where T is the trace of the matter energy–momentum tensor $T_{\mu\nu}$, while ζ is a constant defined by

$$\zeta^{-2} = 6 + \epsilon\xi^{-1} = 6 + 4\omega. \tag{1.14}$$

Notice that φ in (1.13) is the BD scalar field, now given by the combination of ϕ as given by (1.9), though the field equation itself was derived by considering ϕ as an independent field. The fact that the right-hand side of (1.13) is given in terms of the matter energy–momentum tensor guarantees that the force mediated by the scalar field respects the WEP, or universal free-fall (UFF). This is because, in the limit of zero momentum transferred, the source of the scalar field is given by the integrated T_{00}, which is the total energy of the system independent of what the content is.

One might be puzzled to see how the scalar field decoupled from the matter at the level of the Lagrangian comes to have a coupling at the level of the field equations. The underlying mechanism is provided by the nonminimal coupling term which acts as a mixing interaction between the scalar field and the spinless component of the tensor field, as will be elaborated toward the end of Chapter 2.

From (1.13) we expect that the scalar field mediates a long-range force between massive objects in the same way as the Newtonian force does in the weak-field limit of Einstein's gravity. The coupling strength is essentially of the same order of magnitude as that of the Newtonian force as long as ξ or ω is roughly of the order of unity, as we can see by restoring $8\pi G$ in the conventional unit system. Equation (1.14) also shows that the coupling vanishes as $\omega \to \infty$, or $\xi \to 0$. It is often stated that the theory reduces to Einstein's theory in this limit.

According to (1.13) the scalar field does not couple to the photon, for example, indicating that the light-deflection phenomenon will remain unaffected by the scalar force. This is an example displaying how the scalar force makes a difference from general relativity. On re-examining what had been done in general relativity, Brans and Dicke discovered room for the scalar component to be accommodated within the limit $\omega \gtrsim 6$ or $|\xi| \lesssim 0.042$ [5].

Shapiro time-delay measurements during the Viking Project in the 1970s, however, yielded the constraint $\omega \gtrsim 1000$, or $\xi \lesssim 2.5 \times 10^{-4}$ [6]. Two decades later, the latest bound from the VLBI experiments, basically concerning the light-deflection phenomenon involving light from extragalactic radio sources, is even stronger [7]:

$$\omega \gtrsim 3.6 \times 10^3, \quad \text{or} \quad \xi \lesssim 7.0 \times 10^{-5}, \tag{1.15}$$

severer than that expected initially by nearly three orders of magnitude. Notice that $\epsilon = +1$ is also implied. There has been no unambiguous evidence for the presence of the additional scalar field, only certain bounds of the parameter having been obtained.

It seems as if Brans and Dicke wished naturally to find positive evidence right at the very beginning. At one time, Dicke, as an experimentalist, wondered whether there was a flaw in comparing theory and observation. He specifically suspected that the Sun is not completely spherically symmetric, which property was used extensively to derive the Schwarzschild solution. He performed his own experiment to re-measure the Sun's oblateness. If the quadrupole moment J_2 turned out to be as large as $\sim 10^{-5}$, as he and Goldenberg reported [8], it would have allowed more deviation from general relativity, resulting in $\omega \sim 5$ or ~ 0.2 for $\epsilon = 1$ or -1, respectively.

Unfortunately, however, subsequent re-measurements by other groups yielded values mostly as small as $\sim 10^{-6}$ for J_2, including the latest even smaller value [7]. In this sense, there seems to be little hope that ω or ξ is close to anywhere around unity. The smallness of ξ has affected considerably the development of the theory during the years that followed. It appeared as if the theory were destined to grow only to occupy an ever smaller territory without an obvious reason, although Dicke himself worked actively on initiating a new era of "experimental relativity." In some sense, the scalar–tensor theory served as an explicit model illustrating what the world could be like if Einstein were not entirely right.

In spite of all these circumstances surrounding the scalar–tensor theory, however, there have been some people who were deeply impressed by the idea that nature's simplest phenomenon, a scalar field, plays a major role, and tried to modify the prototype BD model in such a way that the constraint (1.15) could be evaded. V. Wagoner suggested extending the original model by introducing arbitrary functions of the scalar field, including a mass term as well [9], though without well-defined physical principles to determine those functions.

One of the present authors (Y. F.) proposed that a dilaton, a Nambu–Goldstone (NG) boson of broken scale invariance, might mediate a finite-range gravity (non-Newtonian force) based on an idea in particle physics [10]. O'Hanlon [11], and Acharia and Hogan [12] showed immediately that the dilaton can be identified as the scalar field of a version of the prototype BD model, hence finding that the massive scalar field does have a place in the theory of gravity.

A crucial point in these approaches is that the scalar field is naturally not immune against acquiring a nonzero mass. If the corresponding force-range of the scalar force turns out to be smaller than the size of the solar system, it no longer affects the perihelion advance of Mercury, for example, thus leaving a constraint like (1.15) irrelevant. This will free us from a long-standing curse.

More recently, T. Damour and A. Polyakov showed that extending the prototype BD model in a way allowing the scalar field to enter in a more complicated manner is rather natural from the viewpoint of string theory [13]. They specifically proposed the "least-coupling principle" (LCP) according to which one might be able to understand why the deviation from general relativity is so small if there is any, though the idea is still short of being implemented from a realistic point of view.

In this book we will be interested also in the cosmological constant, which seems to be one of the hot topics at the present time [14, 15]. Although this subject has a long but widely known history, what we face today is quite new, and appears to be a challenge that probably requires something beyond the standard theory. Today might be a time when the discovery of an accelerating universe [14, 15] is in fact a crisis on which physics will thrive [16]. We may more specifically expect this issue to be a fresh ground to which the scalar–tensor theory applies. The situation might provide a chance to go beyond an "alternative theory," suggesting phenomena that had never been thought of in general relativity.

Ideas based on scalar fields have already been attempted, particularly under the name of "quintessence" [17]. Some of these theories are not necessarily related to the scalar–tensor theory. One has more flexibility, but to some extent they are more phenomenological. After giving a brief overview of the recent developments on these subjects, we will see how we reach an understanding of the accelerating universe in terms of the scalar–tensor theory, which has been extended minimally from the prototype BD model, from our point of view. As a further attempt, we apply the theory also to the reported time-dependence of α, the fine-structure constant [18].

As one of the conclusions that has emerged from our efforts, our classical solutions of the cosmological equations are partially chaotic, with very sensitive dependence on the initial values. Closely connected with this is the great likelihood that we are in a transient state before reaching a final, asymptotic state. All these things may alter our traditional view that the universe we see at the present time should have an attractor solution that depends on the initial values supposedly as little as possible. The universe after all might also be like many natural phenomena around us.

1.2 Where does the scalar field come from?

As we stated before, Brans and Dicke assumed that decoupling of the scalar field from the matter part of the Lagrangian occurs. As we see in the following, this is an assumption that hardly seems to be supported by any of the examples of more fundamental theories. They never made it clear how they could avoid this. It appears as if they had never been particularly concerned about whether there were any theories at a deeper level behind their model, which they viewed as an alternative theory in its own right.

It is nevertheless hard to deny that the scalar–tensor theory has attracted wide interest because it appears to provide a small window through which one can look into phenomenological aspects of more fundamental theories to which one is still denied any direct access otherwise.

It is truly remarkable and even surprising to find that a candidate scalar field of the desired nature is provided by the string theory of the late twentieth century, not to mention the KK theory of the 1920s. We will discuss briefly such candidates, starting from a reasonably detailed account of the KK approach. We then move on to the "dilaton" expected from string theory, and further to the recent development of the "brane," which has become a focus of attention even though it is still highly speculative. We also sketch another highly hypothetical idea that is closely related to "noncommutative geometry," which turns out to be yet another supplier of a scalar field.

1.2.1 The scalar field arising from the size of compactified internal space

Shortly after the advent of general relativity, the historic attempts at unification appeared, first due to H. Weyl [19], and then due to Kaluza. Weyl's theory eventually laid the foundation for what was later called gauge theory, the heart of the contemporary version of unification theories, whereas Kaluza's proposal, later known as the KK theory, played a decisive role in making clear the importance of higher-dimensional space-time in string theory, not to mention serving as an ancestor of the scalar–tensor theory.

Kaluza envisioned five-dimensional space-time to which general relativity was applied. One of the spatial dimensions was assumed to be "compactified" to a small circle leaving four-dimensional space-time extended infinitely as we see it. The size of the circle is so small that no phenomena of sufficiently low energies can detect it.

He started with the metric in five dimensions, of which the "off-diagonal" components connecting the four dimensions with the fifth dimension behave as a 4-vector that has been shown to play the role of the electromagnetic potential. In this way the theory offered the unified Einstein–Maxwell theory. The gauge transformation for the potential is interpreted as an isometry transformation along the circle.

The idea was re-discovered later from a more contemporary point of view [20], in particular in connection with the realization that string theory requires higher-dimensional space-time [21]. We outline briefly how the size of compactified internal space behaves as a four-dimensional scalar field precisely of the nature of the prototype BD model, with the parameter determined uniquely in terms of the dimensionality of space-time. See Appendix A for more details of derivations.

Let us assume the "*Ansatz*" for the $D = (4 + n)$-dimensional metric with n-dimensional compactified space:

$$g_{\bar{\mu}\bar{\nu}} = \begin{pmatrix} g_{\mu\nu}(x) & 0 \\ 0 & A(x)^2 \tilde{g}_{\alpha\beta}(\theta) \end{pmatrix}, \tag{1.16}$$

with the radius A, while $\tilde{g}_{\alpha\beta}(\theta)$ means the purely geometrical portion described by the coordinates θ_α with $\alpha, \beta = 1, 2, \ldots, n$. We choose θ_α to be dimensionless, like angles. Notice that we omitted, for the moment, the off-diagonal components for the gauge fields, focusing on the scalar field.

We also have

$$\sqrt{-^{(D)}g} = \sqrt{-g}\, A^n \sqrt{\tilde{g}}, \tag{1.17}$$

where g is the four-dimensional determinant, while $\sqrt{\tilde{g}}$ is related to the volume V_n of compactified space by

$$V_n = A^n \tilde{V}_n, \tag{1.18}$$

where

$$\tilde{V}_n = \int \sqrt{\tilde{g}}\, d^n\theta. \tag{1.19}$$

We then compute the Einstein–Hilbert term in D dimensions, obtaining the effective Lagrangian in four dimensions, by dividing by $\tilde{\mathcal{V}}_n$ for later convenience;

$$\mathcal{L}_4 = \sqrt{-g}L_4, \tag{1.20}$$

where

$$
\begin{aligned}
L_4(x) &= A^n \tilde{\mathcal{V}}_n^{-1} \int \sqrt{\tilde{g}}\frac{1}{2}R\, d^n\theta, \\
&= \frac{1}{8}\frac{n}{n-1}\phi^2 R + \frac{1}{2}g^{\mu\nu}\,\partial_\mu\phi\,\partial_\nu\phi \\
&\quad + \frac{1}{2}\left(\frac{1}{4}\frac{n}{n-1}\phi^2\right)^{1-2/n}\tilde{R},
\end{aligned}
\tag{1.21}
$$

for $n > 1$, with ϕ defined by

$$\phi = 2\sqrt{\frac{n-1}{n}}A^{n/2}. \tag{1.22}$$

It should be noticed that the sign of the second term on the right-hand side of (1.21) shows that the field ϕ is a *ghost*, corresponding to

$$\epsilon = -1, \tag{1.23}$$

in (1.12) [22]. We will show in later chapters, however, that this does not imply physical inconsistencies; the energy of the whole system remains positive due to the mixing interaction with the spinless component of the metric field, a crucial role played by the nonminimal coupling term.

The first term on the right-hand side of (1.21), with R for a curvature scalar in four dimensions, is the nonminimal coupling term shown in (1.12) with a special choice:

$$\xi = \frac{1}{4}\frac{n}{n-1}. \tag{1.24}$$

Notice that this is certainly in conflict with the constraint obtained from the solar-system experiment (1.15), and will be discussed in connection with the idea of spontaneously broken scale invariance.

We further point out that the scalar field enters the nonminimal coupling term always in the form of ϕ^2 in any dimensionality $n > 1$. For the underlying reason, see Appendix A.

In the last term on the right-hand side of (1.21), $\tilde{R}$ is a curvature scalar computed in terms of $\tilde{g}_{\alpha\beta}$, giving a potential of the scalar field unless it gives a cosmological constant for $n = 2$, or $D = 6$. Notice also that no mass term ($\sim \phi^2$) is present.

Obviously (1.21) and (1.22) lose their meaning if $n = 1$, for which there is no kinetic term for the scalar field. We simply choose

$$\phi = \xi^{-1/2} A, \tag{1.25}$$

with $\epsilon = 0$ and $\xi > 0$ in (1.12). Also we have $\tilde{R} = 0$ for $n = 1$. We do not use (1.10), but (1.13) is still obtained with $\zeta^2 = \frac{1}{6}$, which agrees with what we find by using $\epsilon = 0$ in (1.14). This shows that ϕ still has a dynamical degree of freedom in spite of the absence of the kinetic term. This is again due to the nonminimal coupling term that induces mixing with the tensor gravitational field.

It might be relevant here to add a remark in passing on the absence of the kinetic term of the scalar field if $D = 5$. This is in an apparent contradiction with (1.1) proposed by Jordan, who based his conclusion on the analysis of his five-dimensional theory. This indicates that he dealt with something that is not exactly the same as what we understand as the KK theory with a recipe for compactification, as we find today. It nonetheless still seems true that he was greatly inspired by the KK theory.

We point out that the nonminimal coupling term arises because of the occurrence of A in the determinant, as given by (1.17). This indicates that the scalar field should likely appear in the matter Lagrangian as well, unless there is a special mechanism to prohibit it. From this point of view, it seems difficult for one of the BD requirements to be implemented in the KK approach.

1.2.2 The dilaton from string theory

We find a new breed of scalar fields emerging from string theory, which has been the focus of extensive studies for the past few decades, as one of the most promising approaches toward unification.

It was shown that a closed string has a zero mode described by a symmetric second-rank tensor that behaves in the low-energy limit like the space-time metric. This was done by demonstrating that the interaction among strings occurs in the same way as that in which gravitons interact with each other according to general relativity. It was also shown that the graviton in this context has companions, a scalar field Φ, coming from the trace of a symmetric second-rank tensor, and an antisymmetric second-rank tensor field $B_{\bar{\mu}\bar{\nu}}$.

The field equations of these zero-mode fields were derived explicitly as shown in Eq. (3.4.56) in [23]:

$$R_{\bar{\mu}\bar{\nu}} = -2\nabla_{\bar{\mu}}\nabla_{\bar{\nu}}\Phi + \tfrac{1}{4}H_{\bar{\mu}\bar{\rho}\bar{\sigma}}H_{\bar{\nu}}^{\bar{\rho}\bar{\sigma}}, \tag{1.26}$$

$$\nabla_{\bar{\lambda}}H^{\bar{\lambda}\bar{\mu}\bar{\nu}} - 2\left(\partial_{\bar{\lambda}}\Phi\right)H^{\bar{\lambda}\bar{\mu}\bar{\nu}} = 0, \tag{1.27}$$

$$R = 4\left(\Box\Phi + (\partial\Phi)^2\right) + \frac{1}{12}(HH), \tag{1.28}$$

re-expressed according to our own sign convention, with an obvious notation (HH). We are in the critical dimension $D = 10$ or 26, depending on whether supersymmetry is included or not, respectively. Also the totally antisymmetric field strength is defined by

$$H_{\bar{\mu}\bar{\nu}\bar{\lambda}} = \partial_{\bar{\mu}}B_{\bar{\nu}\bar{\lambda}} + \text{cyclic permutation}, \tag{1.29}$$

while $H^{\bar{\mu}\bar{\nu}\bar{\lambda}}$ is given by raising indices with the help of the inverse metric in D dimensions.

The equations (1.26)–(1.28) can be shown to be derived from the Lagrangian;

$$\mathcal{L}_{\text{st}} = \sqrt{-g}e^{-2\Phi}\left(\frac{1}{2}R + 2g^{\bar{\mu}\bar{\nu}}\,\partial_{\bar{\mu}}\Phi\,\partial_{\bar{\nu}}\Phi - \frac{1}{12}H_{\bar{\mu}\bar{\nu}\bar{\lambda}}H^{\bar{\mu}\bar{\nu}\bar{\lambda}}\right). \tag{1.30}$$

The second term on the right-hand side resembles the kinetic term of Φ, except for the multiplicative factor $e^{-2\Phi}$. This term can be converted into the standard kinetic term if we introduce the field ϕ by putting

$$\phi = 2e^{-\Phi}, \tag{1.31}$$

as one can easily show by recognizing that $d\phi = -2e^{-\Phi}\,d\Phi$. In fact we re-express (1.30) as

$$\mathcal{L}_{\text{st}} = \sqrt{-g}\left(\frac{1}{2}\xi\phi^2 R - \frac{1}{2}\epsilon g^{\bar{\mu}\bar{\nu}}\,\partial_{\bar{\mu}}\phi\,\partial_{\bar{\nu}}\phi - \frac{1}{24}\xi\phi^2(HH)\right), \tag{1.32}$$

where

$$\xi^{-1} = 4 \tag{1.33}$$

and

$$\epsilon = -1. \tag{1.34}$$

Here the first two terms on the right-hand side of (1.32) agree with the D-dimensional version of the BD Lagrangian (1.12) [24, 25]. The factor

ζ^{-2} defined in (1.14) in four dimensions is replaced by its D-dimensional version given by (6.36):

$$\zeta_D^{-2} = 4\frac{D-1}{D-2} + \epsilon\xi^{-1}. \tag{1.35}$$

For $\epsilon\xi^{-1} = -4$, this reduces to

$$\zeta_D^{-2} = \frac{4}{D-2}, \tag{1.36}$$

which is always positive for any $D > 2$, satisfying the condition that the "diagonalized" scalar field be a nonghost field.

Suppose that we are simple-minded enough to expect that each of the $D - 4$ dimensions is compactified trivially with a common radius $A = \text{constant}$. Then (1.32), except for the last term for the moment, reduces to the four-dimensional Lagrangian (1.12). Positivity for the physical mode is assured, but the result $\xi = \frac{1}{4}$, (1.33), is in contradiction with the observational constraint (1.15), as shown in the model of the KK theory, suggesting again the need for a departure from the original model.

We point out, however, that it is still not entirely clear how one can go to four dimensions from D dimensions. There might be some other way of compactification by which the effective coefficient ξ in four dimensions can be made sufficiently smaller than the "original" $\xi = \frac{1}{4}$.

Before going into any such details, one might ask what the underlying reason for the prototype BD model to have emerged from string theory is. The question may be traced back to why the overall common factor $e^{-2\Phi}$ appeared in (1.30). The answer will be given below.

String theory offers a way to avoid the divergences that have plagued traditional field theory for decades. Obviously this is primarily because a string is an extended object, which is contrary to the concept of point particles upon which field theory is based. From a more technical point of view, however, finiteness is due to an invariance under conformal transformation in two-dimensional space-time in which propagating strings reside. Moreover, one has to protect this classical invariance from being broken by quantum effects. A fully consistent theory of strings can be obtained only if what are known as quantum anomalies are removed. In fact, the field equations we mentioned before can be derived also as anomaly-free conditions.

The two-dimensional conformal invariance has its descendant in D-dimensional field theory, namely dilatation invariance, which is implemented with the help of the scalar field in precisely the same way as in the prototype BD model. The relevant transformations are simple extensions of (1.73) and (1.74);

$$g_{\bar{\mu}\bar{\nu}} \rightarrow g_{*\bar{\mu}\bar{\nu}} = \Omega^2 g_{\bar{\mu}\bar{\nu}}, \tag{1.37}$$

$$\phi \rightarrow \phi_* = \Omega^{1-d}\phi, \tag{1.38}$$

$$B_{\bar{\mu}\bar{\nu}} \rightarrow B_{*\bar{\mu}\bar{\nu}} = \Omega^2 B_{\bar{\mu}\bar{\nu}}, \tag{1.39}$$

with $d = D/2$, where we restricted ourselves to $\Omega = $ constant.

Under (1.37), we find

$$\sqrt{-g}R = \Omega^{2-D}\sqrt{-g_*}R_*, \tag{1.40}$$

which implies that the Einstein–Hilbert term is *not* invariant unless $D = 2$.

Also, comparing (1.38) and (1.39) reveals that the present transformation is different from the scale transformation involving the mass dimension of fields, as was mentioned before. The field ϕ or Φ is naturally called a *dilaton*.

Another point to be discussed is the question of whether the dilaton couples to matter fields, like gauge fields and fermions, at the level of the Lagrangian. The same argument that led to dilatation invariance on the zero modes of closed strings likely applies to other matter fields coming from open strings, hence yielding direct dilaton–matter coupling. According to the analyses available so far, it does not seem that there is any simple way to forbid this coupling to matter, though details have yet to be worked out. In this sense one might have to accept the possibility of violation of the WEP as one of the natural consequences of string theory as well as of the KK model. This was what Jordan must have realized from the KK approach, was also noted by Fienz, but was rejected by Brans and Dicke, who chose to appreciate the validity of the WEP as an empirical fact. We also suggest that the coupling to matter can be closely connected with the mechanism by which the massless dilaton acquires a nonzero mass, as shown by an explicit model to be elaborated in Chapter 6.

As we saw above, a major limitation to this approach lies in complications and ambiguities in implementing realistic compactification to four-dimensional space-time. For this reason, we are still not sure whether the simple scale invariance like (1.37)–(1.39) descends to four dimensions. We have no reliable way of knowing how large the effects of violation of the WEP should be either.

For the same reason, we have no way to reject the scenario of the LCP due to Damour and Polyakov [13], who tried to derive a model of the scalar–tensor theory in which the nonminimal coupling term is more complicated than ϕ^2 as in the prototype BD model. They expect instead a function $F(\phi)$ that has a maximum at $\phi = \phi_1$, a certain constant.

Since this maximum serves eventually as a minimum in the potential, one expects that ϕ tends to ϕ_1, for which ϕ is decoupled from the rest of the system, thus leaving the world of general relativity. In the present epoch we are coming close to this asymptotic state, having only a small amount of time-variation of coupling constants and WEP-violating phenomena.

As we are going to show later, on the other hand, we raise the possibility that we are not in the asymptotic state, but still in a transient state, which allows an understanding of the accelerating universe and a time-dependent fine-structure constant, for example. We accept the prototype BD model with certain modifications as an "empirical" rule, which may well not allow immediate derivation from string theory, but offers a suggestion regarding the fundamental theory of what the desired compactification should be like.

1.2.3 The scalar field in a brane world

String theory predicts a new type of nonlinear structure, which is called a *brane*, a nomenclature created artificially from "membrane" [26]. It is a boundary layer on which edges of open strings stand. The idea of a brane with duality plays an important role in the statistical derivation of the entropy of a black hole. This also suggests a new perspective in cosmology; we are living in a brane world, which is a three-dimensional hypersurface in a higher-dimensional space-time [27]. In contrast to the already familiar notion that we live in four-dimensional space-time with n dimensions compactified as an "internal space," which we inherited from the KK theory, our worldview appears to be changed completely. In the following, we are going to review briefly what we should expect to see, though it is not entirely clear for the moment how this picture is related to the scalar–tensor theory.

According to the results of recent progress in superstring theory, different string theories are connected with each other via dualities, making them unified to the *M-theory* in 11 dimensions [28]. Among string theories, the ten-dimensional $E_8 \times E_8$ heterotic string theory is a strong candidate for our real world because the theory may contain the standard model. Hořava and Witten showed that this heterotic string model is equivalent to an 11-dimensional realization of M-theory compactified on the orbifold $R^{10} \times S^1/Z_2$ [29]. Each gauge field on E_8 is confined to the ten-dimensional boundary brane of S^1/Z_2. The ten-dimensional space-time is compactified to $M^4 \times CY^6$, where M^4 and CY^6 are four-dimensional Minkowski space-time (our universe) and Calabi–Yau space, respectively. Particles in the standard model are expected to be confined to this four-dimensional Minkowski space-time, whereas the gravitons propagate in the entire *bulk* space-time.

A new type of KK cosmology based on this brane world picture was proposed by Arkani-Hamed *et al.* [30]. Ordinary matter fields are confined on the brane which is infinitesimally "thin" mathematically, though it may be thick physically, probably of the order of the Planck length. Compared with this thickness, on the other hand, the extra dimensions where only gravitons propagate could be larger. How large it is can be ascertained only by gravitational experiments [31]. Since all of the experiments performed to confirm Newtonian gravity have been carried out above the 1-mm scale, the laws of gravity might be different only below this scale. We re-emphasize that the extra-dimensional space was a tiny internal space of matter fields in the KK approach, but is now a larger space wherein only gravity resides.

Suppose that the fundamental theory of gravity is given by the Einstein–Hilbert term in $D (= 4 + n)$ dimensions and the entire space-time is compactified into four-dimensional space-time times n extra dimensions. This scenario might be formulated by writing an equation for the action:

$$
\begin{aligned}
S &= \frac{1}{16\pi G_D} \int d^D x \sqrt{-^{(D)}g} R \\
&= \frac{1}{16\pi G_D} V_n \int d^4 x \sqrt{-g} R = \frac{1}{16\pi G} \int d^4 x \sqrt{-g} R,
\end{aligned}
\tag{1.41}
$$

where V_n is the volume of extra dimensions given by (1.18), while G_D is a gravitational constant in D-dimensional space-time. In an analogy with (1.2) we may define the D-dimensional Planck mass $M_P^{(D)}$ by

$$
8\pi G_D = \left(M_P^{(D)} \right)^{-(n+2)}.
\tag{1.42}
$$

Notice that, in D dimensions, G_D^{-1} has mass dimension $D - 2 = n + 2$. Using this in (1.41) we obtain

$$
(M_P)^2 = \left(M_P^{(D)} \right)^{(n+2)} V_n.
\tag{1.43}
$$

According to the old KK idea, the size of extra dimensions should be much smaller than 10^{-17} cm ($\sim 1 \, \text{TeV}^{-1}$), so that they remain undetected by low-energy experiments. We have even assumed that the size is nearly as small as the Planck length, the inverse of M_P. In the language of $M_P^{(D)}$, this corresponds to the choice $M_P^{(D)} = M_P$.

If we believe, on the other hand, that ordinary matter fields are confined on a brane world, the extra dimensions are not necessarily required to be so small. This might also be connected with a conjecture that the mass scale $M_P^{(D)}$ in D-dimensional space-time at the more fundamental level is as low as $\sim 1\,\mathrm{TeV}$, nearly the same as the electroweak mass scale. This is expected to remove what is called a "hierarchy problem," which has loomed all the time whenever we try to understand a huge difference between M_P and the energy scale of particle physics in the usual sense. In other words, the hierarchy problem is now interpreted as being present only because we live in four-dimensional space-time which is so "distant" from the fundamental space-time in D dimensions.

Once we accept this idea, we use (1.43) to derive a typical size r_0 of the extra dimensions in the following way:

$$r_0 \sim (V_n)^{1/n} \sim 10^{(30/n)-17}\,\mathrm{cm}, \qquad (1.44)$$

where we have used the values given by (1.3).

If $n = 1$, we expect $r_0 \sim 10^{13}\,\mathrm{cm}$ (~ 1 astronomical unit), which is not possible. If $n = 2$, on the other hand, we find $r_0 \sim 1$ mm, precisely the shortest distance only above which the Newtonian inverse-square law has been tested to within a certain precision, as stated before. In addition to this interesting possibility with the unification scale within reach of our experiments in the near future, also still in accordance basically with the KK approach, there is another alternative scenario based on the brane world picture.

According to the proposal made by Randall and Sundrum [32, 33], the Hořava–Witten model just mentioned can be simplified to a five-dimensional theory in which matter fields are confined to four-dimensional space-time while gravity acts in five dimensions, because the six-dimensional Calabi–Yau space is smaller by at least an order of magnitude than the remaining five-dimensional space-time [34]. They simplified the model further by assuming that our brane is identical to a domain wall in five-dimensional anti-de Sitter space-time with a negative cosmological constant Λ. The five-dimensional space-time is described by the metric, which is not factorizable:

$$ds^2 = e^{-2|y|/\ell} g_{\mu\nu}(x)\, dx^\mu\, dx^\nu + dy^2, \qquad (1.45)$$

where $\ell = \sqrt{-6/\Lambda}$. To find the Minkowski space on the brane ($g_{\mu\nu} = \eta_{\mu\nu}$), the tension $\mathcal{T}$ of the brane must satisfy $|\mathcal{T}| = 3/(4\pi\ell G_5)$. The "warp" factor $e^{-2|y|/\ell}$ which is rapidly changing in the extra dimension plays a very important role, in contrast to the usual KK compactification. They discussed two models.

In their first model (the Randall–Sundrum type-I model) [32], they proposed a mechanism to solve the hierarchy problem by incorporating a small extra dimension bounded by two boundary branes, with positive and negative tension, located at $y = 0$ and $y = s$, respectively, as illustrated in Fig. 1.1.

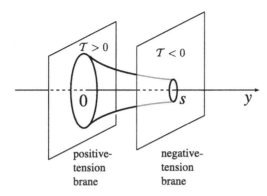

Fig. 1.1. The Randall–Sundrum type-I model. Since gravity is confined on the brane $\mathcal{B}$, it can be described by the induced metric of $\mathcal{B}$. The circles describe the warp factor $e^{-2|y|/\ell}$.

By assuming that we are living in a negative-tension brane, we find a solution to the hierarchy problem. We first estimate the four-dimensional Planck mass scale by integrating the five-dimensional action in the fifth direction as

$$S_g^{\rm eff} = \int d^4x \int_0^s dy \frac{1}{16\pi G_5} \sqrt{-^{(5)}g}\, ^{(5)}R$$
$$\sim \frac{1}{16\pi G_5} \int_0^s dy e^{-2|y|/\ell} \cdot \int d^4x \sqrt{-g} R, \qquad (1.46)$$

where we made an approximate estimate $^{(5)}R \sim e^{2|y|/\ell}R$, without solving Einstein's equation in five dimensions rigorously. Note also that the separated fifth dimension has no curvature. We then find

$$M_{\rm P}^2 = \frac{1}{8\pi G} \sim \frac{1}{8\pi G_5} \int_0^s dy\, e^{-2|y|/\ell} = \frac{\ell}{16\pi G_5}\left(1 - e^{-2s/\ell}\right). \qquad (1.47)$$

This means that $M_{\rm P}$ depends weakly on the distance s, as long as $e^{-s/\ell} \ll 1$. To show how the hierarchy problem is resolved, we consider a fundamental Higgs field confined in the visible brane with negative tension. The action is given by

$$S_{\rm vis} = \int d^4x \sqrt{-g_{\rm vis}}\left[g_{\rm vis}^{\mu\nu}D_\mu H^\dagger D_\nu H - \lambda\left(|H|^2 - v_0^2\right)^2\right], \qquad (1.48)$$

where $g_{\mu\nu}^{\rm vis}$ denotes the four-dimensional components of the five-dimensional metric evaluated at $y = s$, i.e. $g_{\mu\nu}^{\rm vis} = e^{-2s/\ell}g_{\mu\nu}$. This, together with redefinition of the Higgs field, $H \to e^{s/\ell}H$, leads to

$$S_{\rm vis}^{\rm eff} = \int d^4x \sqrt{-g}\left[g^{\mu\nu}D_\mu H^\dagger D_\nu H - \lambda\left(|H|^2 - v_{\rm eff}^2\right)^2\right], \qquad (1.49)$$

where $v_{\text{eff}} = e^{-s/\ell}v_0$ gives the physical symmetry-breaking energy scale, which could be much smaller than the original energy scale v_0. In fact, if $s/\ell \sim 35$, this produces a TeV energy scale from the four-dimensional Planck scale M_P. This may be a natural explanation of the hierarchy problem. In this discussion, although we assumed that $M_P^{(5)} = (8\pi G_5)^{-1/3} \approx M_P$ is the fundamental energy scale, we are allowed to consider that the TeV scale is fundamental and the Planck scale is induced, contrary to the conventional wisdom.

In their second model (the Randall–Sundrum type-II model) [33], on the other hand, we assume that we are living in the positive-tension brane surrounded by AdS. There is no second brane with negative tension, which is obtained from a two-brane model in the limit of $s \to \infty$. Although hierarchy is still left unsolved, four-dimensional Newtonian gravity is recovered at low energies even if the extra dimension is not compact. This is proved by applying a perturbation approximation to the above solution (1.45) with a positive-tension Minkowski brane at $y = 0$. Consider perturbation to the four-dimensional components, $^{(5)}g_{\mu\nu} = e^{-2|y|/\ell}\eta_{\mu\nu} + h_{\mu\nu}$. By setting $h(x, y) = \hat{\psi}(z)e^{-|y|/(2\ell)}e^{ipx}$ with $z = \ell(e^{|y|/\ell} - 1)$ and $p^2 = -m^2$, we find the perturbation equation of the graviton to be

$$\left[-\tfrac{1}{2}\partial_z^2 + V(z)\right]\hat{\psi} = m^2\hat{\psi}, \qquad (1.50)$$

where

$$V(z) = \frac{15}{8\ell^2(|z|/\ell + 1)} - \frac{3}{2\ell}\delta(z) \qquad (1.51)$$

is a volcano-shaped potential. Note that the indices of metric perturbations are the same in all terms when one is working in the gauge of $\partial^\mu h_{\mu\nu} = h_\mu{}^\mu = 0$. The volcano-shaped potential confines a massless mode on the brane. As a result, even if the fifth dimension is not compact, the Newtonian gravitational potential is recovered in the low-energy limit as

$$V(r) \sim G\frac{m_1 m_2}{r}\left(1 + \frac{\ell^2}{r^2}\right), \qquad (1.52)$$

where m_1 and m_2 are masses of two particles on the brane. This "compactification" is completely different from the KK-type compactification.

Although there are many interesting ideas in this new field, we will focus here on the possibility of the occurrence of a scalar field in the brane world. The scalar field may arise in two different forms, either the one similar to ordinary moduli fields associated with the process of compactification, such as in the manner of the conventional KK approach and in superstring models, or the one which is unique to the models of the brane world. The scalar field of the latter type corresponds to a "distance" between two branes, often called a *radion*.

In the first model due to Randall and Sundrum, for example, there are two branes which are the boundaries of the fifth dimension S^1/Z_2. In order to understand the hierarchy problem, the distance must be related to the ratio of two extremely different scales (Planck and electroweak scales), but it can remain arbitrary because no force acts on the two branes. This implies that the distance plays the role of a kind of massless scalar field in our four-dimensional world. In terms of linear perturbations of the Randall–Sundrum model, Garriga and Tanaka showed that the effective theory of gravity in four dimensions is in fact the prototype BD model with the parameter

$$\omega = \tfrac{3}{2}\left(e^{\pm s/\ell} - 1\right),\tag{1.53}$$

where $\pm$ corresponds to the sign of the tension of branes [35]. Since we have to live in a negative-tension brane in order to solve the hierarchy problem, ω is negative but still larger than $-\tfrac{3}{2}$. According to (1.10), we find $\epsilon = -1$ and $\xi^{-1} < 6$ in our notation. Also the positivity condition for (1.14) implies that the diagonalized field σ is not a ghost, in spite of the apparent contrary conclusion because of $\epsilon < 0$. Interestingly enough this feature is shared with the dilaton expected in the basic string theory in D dimensions, and many other examples. See also the remark in subsection 1.3.2.

In the second Randall–Sundrum model, this new type of scalar field no longer appears. When we reduce it to a four-dimensional effective theory, however, we find two new terms in the extended versions of Einstein's equations, which will change the dynamics of four-dimensional space-time and fields. We will discuss them in Chapter 5.

In the context of string theory or M-theory, one would also expect scalar fields associated with many moduli fields in the gravitational sector, which are allowed, in principle, also to propagate in the bulk, by which we mean the entire five-dimensional space-time including one dimension that separates the two branes [34, 36, 37]. Lukas, Ovrut, and Waldram [34], for example, derived an effective five-dimensional action by dimensional reduction from 11-dimensional M-theory.

Somewhat like the way in which a scalar field arises from the size of internal space in KK theory as explained in subsection 1.2.1, the volume V of five-dimensional bulk yields the dilaton field ϕ by the relation

$$V \equiv \exp(-\sqrt{2}\phi)\tag{1.54}$$

in (1.42) for $D = 5$. The part of the action in which ϕ occurs is given by

$$S = \int_{\mathcal{M}_5} d^5x \sqrt{-^{(5)}g} \left(\frac{1}{2} {}^{(5)}R - \frac{1}{2} (\nabla\phi)^2 - \frac{K^2}{6} e^{-2\sqrt{2}\phi} \right)$$
$$+ \sqrt{2}K \int_{\mathcal{B}^{(1)}} d^4x \sqrt{-g} \, e^{-\sqrt{2}\phi} - \sqrt{2}K \int_{\mathcal{B}^{(2)}} d^4x \sqrt{-g} \, e^{-\sqrt{2}\phi}, \quad (1.55)$$

where K is a constant, while ${}^{(5)}g$ and ${}^{(5)}R$ stand for the determinant and scalar curvature in five dimensions, respectively. Also $\mathcal{M}_5$ is five-dimensional bulk space-time with a positive definite potential, while $\mathcal{B}^{(1)}$ and $\mathcal{B}^{(2)}$ represent two four-dimensional boundary branes with negative and positive tensions, respectively.

The five-dimensional scalar field ϕ appears also in our four-dimensional world, but is different from a purely four-dimensional object, because it is affected by the behavior of a field living in five dimensions.

1.2.4 The scalar field in the assumed two-sheeted structure of space-time

As yet another focus of recent intensive studies on unification, we mention noncommutative geometry. Of particular interest from our point of view is that this approach provides scalar fields as gauge fields in *discrete* spaces introduced in addition to ordinary continuous space-time. These scalar fields may be identified, in the electroweak unified theory, for example, with the Higgs fields and the desired potentials. T. Saito and his collaborators moved ahead to show that a natural extension to the theory of gravity leads to a scalar–tensor theory [38]. Interestingly, however, they did this without using the technique of noncommutative geometry, but relied on the more conventional analysis of differential geometry, starting with the assumption that our four-dimensional space-time is two-sheeted, as implemented by five-dimensional space-time with the structure $M_4 \times Z_2$. As will be sketched below, this is one of the rare explicit models in which the scalar field appears as a normal (nonghost) field, unlike in the KK thoery and in string theory. See Appendix B for more details.

As in KK theory, let us introduce five coordinates $x^{\bar{\mu}}$, where x^μ for $\mu = 0, \ldots, 3$ are those in the four-dimensional manifold M_4, while x^4 takes two integer values, say x^4_+ and x^4_-, corresponding to which of the two M_4's we are in. One might interpret this as meaning that one of the five coordinates is discretized spontaneously, resulting in degenerated two-sheeted space-time. There is an operation called r that takes us from one of the M_4's to another. Obviously, performing two successive operations takes us back to the original M_4, represented symbolically by

$$x^4 + r + r = x^4. \quad (1.56)$$

On the other hand, we assume that the varieties of physics on the two M_4's should be equivalent to each other. In order to implement this condition, we introduce a continuous variable $w(x^4)$, such that the difference

$$\Delta w(x^4) = w(x^4 + r) - w(x^4) \tag{1.57}$$

is the distance between the two M_4's. We choose

$$\Delta w(g) \to 0, \tag{1.58}$$

corresponding to the situation that the two space-times are pasted onto each other. Using (1.56), we find

$$w(x^4 + r + r) = w(x^4). \tag{1.59}$$

Owing to this property, any function $F(x, x^4)$ is expressed as

$$F(x, x^4) = F_0(x) + F_1(x) w(x^4), \tag{1.60}$$

which implies that a Taylor expansion with respect to $w(x^4)$ terminates at the linear order.

Differentiation ∂_4 with respect to x^4 may be defined by ∂_w with respect to the continuous variable w, but with a subtle point that x^4 has only two values. An example is illustrated by the following:

$$
\begin{aligned}
\partial_w \Delta w(x^4) &= \frac{\Delta w(x^4 + r) - \Delta w(x^4)}{w(x^4 + r) - w(x^4)} \\
&= \frac{[w(x^4 + r + r) - w(x^4 + r)] - [w(x^4 + r) - w(x^4)]}{\Delta w(x^4)} \\
&= \frac{2w(x^4) - 2w(x^4 + r)}{\Delta w(x^4)} = -2,
\end{aligned}
\tag{1.61}
$$

where we used (1.57) and (1.59).

By developing the analysis on the parallel transport we find

$$\partial_w g_{\bar{\mu}\bar{\nu}} = \Gamma_{\bar{\mu}w\bar{\nu}} + \Gamma_{\bar{\nu}w\bar{\mu}} + g^{\bar{\rho}\bar{\sigma}} \Gamma_{\bar{\mu}w\bar{\rho}} \Gamma_{\bar{\nu}w\bar{\sigma}} \Delta w, \tag{1.62}$$

where $\Gamma_{\bar{\lambda}\bar{\mu}\bar{\nu}} = g_{\bar{\lambda}\bar{\kappa}} \Gamma^{\bar{\kappa}}_{\bar{\mu}\bar{\nu}}$. We find that the coordinate condition can be imposed only in the limit of (1.58).

We then move on to compute curvature tensors, taking subtleties carefully into account, obtaining

$$^{(5)}R = R + g^{\mu\nu} R^4{}_{\mu,4\nu} + g^{44} R^\rho{}_{4,\rho 4}. \tag{1.63}$$

We choose the same *Ansatz* for the five-dimensional metric as that given by (1.16), but with $\mathcal{A} = A^2$ replaced by

$$\mathcal{A}(x, x^4) = \left(A_0(x) + A_1(x) \, \Delta w(x^4) \right)^2, \tag{1.64}$$

in accordance with (1.60). Note that, unlike in the KK approach, the dependence on the "internal coordinate" w does not imply including heavier modes. We also take the limit $\Delta w \to 0$, obtaining

$$\sqrt{-^{(5)}g} = \sqrt{-g}B, \tag{1.65}$$

where

$$B(x) = |A_0(x)|. \tag{1.66}$$

We finally have the five-dimensional Einstein–Hilbert term put into the form

$$\tfrac{1}{2}\sqrt{-g}\,^{(5)}R = \tfrac{1}{2}\sqrt{-g}B \left(R - B^{-2}\nabla_\mu \left(B\nabla^\mu B \right) \right). \tag{1.67}$$

In the second term on the right-hand side we integrate by parts, obtaining

$$\tfrac{1}{2}\sqrt{-g}\partial_\mu \left(B^{-1} \right) \left(B\nabla^\mu B \right) = -\tfrac{1}{2}\sqrt{-g}B^{-1} \left(\partial B \right)^2, \tag{1.68}$$

which can be put into a canonical form

$$-\tfrac{1}{2}\sqrt{-g} \left(\partial\phi \right)^2, \tag{1.69}$$

if we define ϕ by

$$B = \tfrac{1}{8}\phi^2. \tag{1.70}$$

On substituting this into the right-hand side of (1.67) we arrive at the BD Lagrangian (1.12), apart from the matter part, with

$$\xi^{-1} = 8 \tag{1.71}$$

and

$$\epsilon = +1, \tag{1.72}$$

which is particularly in contrast to the previous examples, (1.23) and (1.34).

The reason for having obtained the prototype BD model is now traced back to the exact invariance of the five-dimensional Lagrangian (1.71) under the set of *global* transformations

$$g_{\mu\nu} \to g_{*\mu\nu} = \Omega^2 g_{\mu\nu}, \tag{1.73}$$

$$\phi \to \phi_* = \Omega^{-1}\phi, \tag{1.74}$$

where Ω is a constant.

This time we have a nonvanishing kinetic term even for $n = 1$, and hence the argument is nontrivial, unlike in the five-dimensional KK theory.

Also to be noted is that, as in KK theory, the appearance of $B(x)$ or $\phi(x)$ in the matter Lagrangian seems unavoidable, hence causing violation of the WEP, in general.

1.3 Comments

When we consider the scalar–tensor theory as a spring board for going beyond general relativity, we often feel the necessity of re-examining traditional ideas or concepts from their foundations, or keeping ourselves free from widely accepted presuppositions. This seems true particularly for simple subjects like the WEP and the basic parameter of the prototype BD model. These two subjects will therefore be selected to be discussed first in this section.

The third comment will be made on conformal transformation in order to emphasize its crucial importance for the whole of the discussion of this field.

1.3.1 The weak equivalence principle

Brans and Dicke [5] demanded that the scalar field be decoupled from the matter Lagrangian in order to save the WEP from being violated. This must have been natural because this experimental fact has been supported ever more strongly, by measurements made by Dicke himself [39]. We point out in this connection that the EP as one of the pillars that support the whole structure of general relativity can remain intact even if the WEP is not strictly observed, though it might sound a bit strange. After a series of heuristic arguments starting from a naive WEP, or the law of universal free-fall (UFF), Einstein finally reached a geometrical law, which embodies the EP expressed in a much more abstract manner; a tangential space-time attached to each world point on a Riemannian manifold is Minkowskian. This expression might be called the ultimate equivalence principle (UEP), for convenience in the following discussion.

The WEP is after all a phenomenological law subject to experimental verification, as was demonstrated by the experiments categorized as concerning a fifth force [40]. The physical circumstance that allows violation of the WEP can be described as follows.

Suppose that there is a non-Newtonian force that couples to matter depending on its composition due to the lack of a geometrical nature. The force may also have a finite range. If the strength *happens* to be nearly as great, or weak, as that of gravity, then this force may participate in "gravitational phenomena." Trying to probe any composition-dependence in this type of phenomenon may be considered an attempt to detect a non-Newtonian force. This can be done in a consistent manner in the sense that one can still formulate the theory in a manner that respects general covariance and Riemannian geometry, thus observing the UEP. In this situation one can no longer expect that verifying the WEP automatically verifies the UEP. Gravitational phenomena are unavoidably more complicated, or are simply richer in content.

Suppose that the WEP turns out not to be supported by observation. Does this imply a collapse of Einstein's idea, or anything crucial in fundamental physics? The answer is yes if we take Einstein's theory strictly as the standard theory of general relativity. In fact it has been argued [41] that Einstein's equivalence principle (EEP) is a central law of geometrical space-time, resting entirely on the validity of the WEP, among other things. We may, however, take another attitude in order to place more emphasis on the UEP at a deeper level as the "heart and soul" of Einstein's entire attempt not rigidly connected to the WEP.

This seems to be appropriate in view of the fact that many theoretical models of unification and experimental tests are being attempted. As stated before, from the UEP follows such a powerful mathematical tool as the "comma-to-semicolon rule" [42], which makes it possible to formulate any theory, even with violation of the WEP included, in curved spacetime. We by no means downgrade the importance of the WEP, which applies to the majority of gravitational phenomena, not to mention its heuristic role in the past.

Nearly the same question as above may also be asked about the idea of time-variability of coupling constants. The concept of EEP in the sense of [41] asserts also that the outcome of any local nongravitational experiment is independent of where and when in the universe it is performed, or local position independence (LPI), saying essentially "constant is constant." At present most of the theoretical views which support time dependence are based, explicitly or implicitly, on a theoretical framework in which a slowly varying scalar field enters the process defining the *observed* coupling constants. We may come back eventually to LPI, because we expect to find truly constant coupling parameters behind the observational world. Nevertheless confusions might arise unless the statements are made with considerable care.

A more imminent question may then be raised, namely that of whether one can devise any experiment to test the UEP directly without being bothered by the possibility of there being a non-Newtonian effect. It does not seem easy to answer this in the affirmative. One must take many facts into account, generally speaking. One can go out sufficiently away from the source if the force is known to be of finite range, for example.

It might be useful to remember that the familiar relation $d\tau = \sqrt{-g_{00}}\,dt$ that connects the local time difference dt and the proper time difference $d\tau$, as used in analyzing gravitational redshifts, is a manifestation of the UEP, independent of any details of quantum mechanics that need to be applied to emission and absorption of photons, for example. A closer look at comparison of gravitational redshifts from atoms of H and $\bar{\text{H}}$ can be used as a direct test of the above relation even if a vector- or a scalar-type non-Newtonian force is acting [43].

It might also be useful to quote names for different kinds of equivalence principles. The strong equivalence principle (SEP) means that UFF applies to large objects such as celestial bodies that have considerable amounts of gravitational self-energy [41]. In other words the proportionality between inertial and gravitational masses holds true up to the higher-order effects of gravity. This is an example showing that one does not always know whether the WEP supposed to be established only to the zeroth order serving as a heuristic basis of general relativity continues to be valid in the completed theory [44]. Detailed analysis shows that the SEP is true for general relativity but not for the prototype BD model, which is still a geometrical theory. Experiments have been done on the motion of the Earth–Moon system falling toward the Sun, the "Nordvedt effect" [45], giving a null result so far.

We remember that the same nomenclature had been used before to imply that UFF is valid for general dynamical systems other than simple mechanical systems [46]. Some authors talked about the SEP implying that constant is constant, i.e. in the sense of LPI in the terminology in [41]. In this context, Brans and Dicke called their theory the one which violates the SEP while still maintaining the WEP [5].

The name WEP II is used if objects fall with a common acceleration not only with respect to the center-of-mass trajectory but also with respect to the rotation [47].

In Chapter 2, the WEP is explained first as a geodesic equation of a particle, which is derived if the matter energy–momentum tensor is covariantly conserved; the latter condition is found to be true if the scalar field is decoupled from the matter Lagrangian. However, note that the latter condition is rarely satisfied in many theoretical models of unification, as was mentioned toward the end of section 1.

In later chapters we check whether the WEP holds true for a scalar force, for example, by checking whether the field couples to the trace of the matter energy–momentum tensor, hence verifying the composition-independence of the effect. Needless to say, we accept the validity of this coupling only within the accuracy to which composition-independence has been tested experimentally. With this condition satisfied, we do not have any reason to confine ourselves to the coupling of a scalar field to the trace of the energy–momentum tensor, as is illustrated by explicit examples in Chapters 4 and 5.

It seems likely that the exclusive trace coupling chosen in the prototype BD model might be an over-simplification, though the way the coupling of the scalar field to matter occurs at the level of the Lagrangian should be much more complicated than Jordan proposed.

1.3.2 The value of ω and mass of the scalar field

Rather frequently we hear about the observational constraint (1.15), $\omega \gtrsim 3.6 \times 10^3$ or $\xi \lesssim 7.0 \times 10^{-5}$, obtained from the solar-system experiments. According to (1.10), this also implies that $\epsilon = +1$, or a positive ω. On the other hand, we are going to find many indications of $\epsilon = -1$, which means an apparent ghost nature of the field ϕ. The positive energy, or nonghost nature, of the "diagonalized field" σ is assured if the coupling strength is positive; $\zeta^{-2} = 6 + \epsilon \xi^{-1} > 0$ according to (1.14). This implies that $\xi > \frac{1}{6}$ for $\epsilon = -1$, or $\zeta^2 > \frac{1}{6}$.

We point out moreover that $\epsilon = +1$ is imperative if $\xi < \frac{1}{6}$, as is indicated also by (1.15), in contrast to the opposite sign (1.34) found in string theory. According to the LCP [13], we thus should expect that the function which multiplies the kinetic term of ϕ changes sign owing to the inclusion of higher-order terms. This seems to require a simple reason, if it is not absolutely necessary.

Probably a more natural way to avoid this crucial contradiction between string theory and observation is to expect that the scalar field is massive such that the range of its force is shorter than the size of the solar system in order to make the related experiments not applicable for the present purpose. It seems useful to summarize briefly what the argument is like in order to save the whole idea of the scalar–tensor theory in this direction.

Let us ask ourselves whether there is any theoretical reason for assuming a massless field to start with. In this connection, it seems important to emphasize that the BD Lagrangian given by the first two terms on the right-hand side of (1.12), but with the last term L_{matter} removed for the moment, allows an invariance under the global transformations (1.73) and (1.74), as discussed toward the end of the preceding section.

Equation (1.73) would have been what is known as a conformal transformation if Ω were space-time-dependent, as we will discuss shortly. In this sense (1.73) is its global version, and may be called a *scale transformation*, or *dilatation*, whereas (1.74) is included to complete the invariance.

The Einstein–Hilbert term fails to respect this scale invariance, or dilatation symmetry. The same is true also for the possible mass term of ϕ, if there is one. One would thus be convinced that this invariance is truly unique to the prototype BD model. No extension of the nonminimal coupling term beyond $\xi\phi^2$ is allowed without losing this invariance. This observation is in fact a key to understanding why string theory provides a candidate for the scalar–tensor theory.

In Chapter 6 we discuss how this invariance can be broken first spontaneously, then explicitly, and hence how the scalar field can acquire a nonzero mass. In the first step the scalar field is still a massless NG boson, which is qualified to be called a "dilaton." The second step is triggered by the quantum "anomaly" effect, a natural occurrence due to interactions among matter fields.

This mechanism will be implemented, resulting eventually in non-Newtonian gravity featuring a macroscopic size of the force range, as was already mentioned in the preceding comment on the WEP. The parameter ξ will be determined by experiments of this type, rather than by those on the solar system.

1.3.3 Conformal transformation

As will be discussed in detail in Chapter 3, conformal transformation is important almost inherently in any version of the scalar–tensor theory. As was pointed out first by W. Pauli, one can always eliminate the nonminimal coupling term by this transformation, given by

$$g_{\mu\nu} \rightarrow g_{*\mu\nu} = \Omega^2(x)g_{\mu\nu}, \tag{1.75}$$

with a space-time-dependent $\Omega(x)$ chosen to be $\Omega^2 = \xi\phi^2$. One speaks of moving from one "conformal frame" to another. More specifically, the Lagrangian takes the form of a nonminimal coupling term in the "Jordan frame," or simply the J frame, while in the "Einstein frame," or the E frame after the conformal transformation, the same Lagrangian is re-expressed in terms of the Einstein–Hilbert term. An imminent question is that of which, if any, conformal frame is "physical" enough to be selected.

In reply we first remind readers that conformal transformation is simply *different* from general coordinate transformation; the former changes the line element ds^2 which is kept invariant under the latter.

Secondly, we point out that few theories or models of practical importance are invariant under conformal transformation. As a consequence no notion of invariance or equivalence is useful. In different conformal frames, physical phenomena *look different* in general, though they are related to each other. Whether G is constant or varies with time, for example, depends on which conformal frame we live in.

This reminds us of different non-inertial coordinate systems in Newtonian mechanics. Also the transformation is local. This makes it somewhat difficult to find explicit examples except in the physics of cosmology. Even in this area, it is not always simple to single out a particular conformal frame as physical; sometimes one must rely on overall consistency of the laws, which may depend on the models one uses. This might be a source of the complications or even confusion we often face. A list of related papers can be found in [48].

Although we finally accept that the E frame is physical, or approximately physical, we are still not completely sure about the nature of the J frame. One of the likely answers is that it is a "theoretical" conformal frame. If we want to write a Lagrangian at all, we must choose a conformal frame beforehand. This is true also in Einstein's theory without a scalar field. Generally speaking, one can create any conformal frame one wishes. The only reason why we do not do this is the lack of absolute necessity. One can say, on the other hand, that string theory is formulated in such a way that $e^{-2\Phi}$, as in (1.30), appears as a factor that determines the coupling constant, which we later interpret as a nonminimal coupling term. For this reason the conformal frame is sometimes called a "string frame." It appears that there is no *a priori* reason why this conformal frame is, or is not, physical. The decision should be made only on the basis of what clock or meter-stick we use, as will be discussed in detail in Chapter 3.

We finally add that the J frame has often been accepted as a physical conformal frame because particle masses are chosen to be constant in the prototype BD model. The argument is obviously connected to a natural expectation that the physical standards of length and time are provided basically by the masses of elementary particles. This is, however, precisely what is going to be challenged in cosmology with a cosmological constant added, leading eventually to a revision in Chapter 4, in such a way that particle masses remain constant in the E frame. We emphasize on this occasion that the revision will be made not in the intrinsically gravitational part but in the matter term.

Another related but somewhat vague question might be that of whether the theory in the E frame is indeed simply an "ordinary" theory with a scalar field added. The answer depends on what is meant by ordinary. In fact it is ordinary because there is no nonminimal coupling term, but the matter-gravity interaction is strongly constrained by how the theory was defined in the J frame. An example can be found in the prototype BD model; the lack of coupling to matter at the level of the Lagrangian in the

J frame assures that the coupling in the E frame is restricted in such a way that the WEP is strictly observed. As another example, we quote our own model of the scale-invariant matter–scalar coupling in the J frame, which turns out to imply a complete decoupling in the E frame.

It might be noteworthy that these results can be altered by quantum corrections. Nevertheless, these examples show an interesting feature, namely that certain restrictions in the E frame can be traced back to highly simple rules in the J frame.

Many phenomenological analyses, in terms of quintessence, for example, are made with the standard Einstein–Hilbert term, rather freely without being much constrained by the scalar–tensor theory. It might be a good point, however, to try to see what one could see if one were to move "back" to the J frame. This would be a nontrivial task because one does not know beforehand what the J frame is, or even whether it uniquely exists at all.

1.3.4 Mach's principle and Dirac's suggestion for time-dependent G

It is widely known that Mach's principle and Dirac's suggestion for time-dependent G were behind the advent of the scalar–tensor theory. Let us summarize today's view on these ideas.

When Jordan noticed that a four-dimensional scalar field can emerge from five-dimensional theories either in the approach from projective geometry or in the theory due to Kaluza and Klein, he was obviously encouraged by the possibility of a time-dependence of G. He wished to find out whether the idea can be implemented by invoking a dynamical scalar field without sacrificing the bases of the special and general theories of relativity. As it turned out, however, the expected time-dependence $G \sim t^{-1}$ has never been obtained in the cosmological solution of the scalar field, in accordance with other observational constraints, as will be found in Chapter 4.

More crucial from a theoretical point of view, however, is that the calculated time-dependence depends on the conformal frame one chooses. An extreme example is what is called the Einstein frame, in which there is no nonminimal coupling term, with a purely constant G. The question of which conformal frame is physical depends on the model, which will be one of the main themes in Chapters 4 and 5. Depending on the situation, G may depend on time, but in a way quite different from the manner Dirac suggested.

In this connection we recall that Dirac's argument was based originally on his *large-numbers hypothesis* (LNH). According to him, one of the best known examples of a very large (or small) number is the ratio between the two coupling constants:

$$\mathcal{N} \equiv \frac{Gm_e^2}{\alpha} = \frac{1}{8\pi\alpha}\left(\frac{m_e}{M_P}\right)^2 \approx 2.31 \times 10^{-43}, \tag{1.76}$$

where m_e is the electron mass while $\alpha = e^2/(4\pi\hbar c) \sim 1/137$ is the fine-structure constant. Dirac thought that ratios between fundamental constants should be of the order of unity. From this point of view, the number $\mathcal{N}$ must involve some large number that is not of really fundamental nature.

Suppose that G is not a true constant but varies as the inverse of the age of the universe:

$$G \sim t^{-1}, \tag{1.77}$$

where t is the cosmic time. It then follows that

$$\frac{G(t)}{G(t_0)} = \frac{t_0}{t}, \tag{1.78}$$

where t_0 is the present age of the universe. The value $t_0 \approx 1.3 \times 10^{10}$ years corresponds to 1.26×10^{60} in units of the Planck time, as given before. Instead of this, let us consider a fundamental time such as the time t_e given by

$$t_e = \frac{\hbar}{m_e c^2} \approx 4.9 \times 10^{21}, \tag{1.79}$$

again in units of the Planck time. From (1.78) it then follows that

$$\frac{G(t_e)}{G(t_0)} \approx 2.59 \times 10^{38}.$$

Substituting this into (1.76), and assuming the constancy of m_e and α, we finally obtain

$$\mathcal{N}(t_e) \approx 0.60 \times 10^{-4}, \tag{1.80}$$

which might be considered of order unity. The smallness of $\mathcal{N}$ at the present time is simply because our universe is old. This attractive view does not appear to be supported by subsequent developments of the scalar–tensor theory, as we mentioned.

It might be worth trying to see whether the right-hand side of (1.80) can be made closer to unity by changing the right-hand side of (1.77) to t^{-p}. An easy exercise yields $p = 1.11$. This implies that the LNH continues to give a desired number only if the power $-p$ is "fine-tuned" to -1 approximately to within an inaccuracy of 10%. We find, on the other hand, the following observational upper bounds on $\dot{G}/G$:

$$\frac{\dot{G}}{G} = \begin{cases} (0.2 \pm 0.4) \times 10^{-11}\,\text{years}^{-1} & \text{(Viking Project [49]),} \\ (-0.06 \pm 0.2) \times 10^{-11}\,\text{years}^{-1} & \text{(binary pulsar [50]),} \end{cases}$$

which is already below $\sim 10^{-10}$ years^{-1} by an order of magnitude, but may still appear to be roughly consistent with Dirac's prediction. We show, however, that this is not the case.

Suppose that we approximate the upper bounds by

$$\dot{G}/G = -10^{-11} \text{ years}^{-1} \approx -0.1 \times t_0^{-1}, \tag{1.81}$$

where we have discarded the choice of an increasing $G(t)$. (1.81) implies that

$$G(t) \sim t^{-0.1}. \tag{1.82}$$

The power -0.1 is obviously outside the range for which the LNH works. In fact we now have $G(t_{\mathrm{e}})/G(t_0) \approx 6.9 \times 10^3$, hence giving $\mathcal{N}(t_{\mathrm{e}}) \approx 1.6 \times 10^{-39}$ instead of the right-hand side of (1.80). For this reason we judge that the LNH is now completely ruled out.

Is there then any alternative way to understand very large or small numbers in nature? The relation (1.76) may be interpreted as giving

$$\frac{m_{\mathrm{e}}^2}{M_{\mathrm{P}}^2} \approx 5.78 \times 10^{-42} \alpha \approx 4.26 \times 10^{-44}.$$

This may be re-expressed as

$$\frac{m_{\mathrm{e}}}{M_{\mathrm{P}}} \approx \exp\left(-\frac{b}{2\alpha}\right), \tag{1.83}$$

where the number b is estimated to be ~ 0.74, certainly of the order of unity. This kind of nonperturbative dependence on a coupling constant has also been suggested [51, 52] without invoking the age of the universe. We also point out that the same type of dependence is suggested by the renormalization-group equation [53]. More recently, another solution of this "hierarchy problem," also in terms of an exponential dependence, has been suggested in the brane cosmology, as will be explained in Chapter 5.

These arguments do not require assuming time-variability of the coupling constants. Quite apart from the context of the LNH, however, we may still consider time-variation of constants, now from the point of view of unified theories. This is because the observed *effective* coupling constants might be derived from time-independent coupling constants at the more fundamental level of the theories. It is also reasonable to expect that this process of derivation comes with the participation of scalar fields, which vary slowly as the universe evolves. It is quite likely that the resulting effective coupling constants vary in ways totally different from the manner Dirac expected. An explicit example with the "dilaton" as the relevant scalar field will be shown in Chapter 6. This example is unique also in the sense that the time-dependence of α is related to the accelerating universe, a contemporary version of the problem of the cosmological constant.

Another issue that historically influenced the pioneers of the scalar–tensor theory strongly is Mach's principle. They focused on the particular expression

$$G \sim R_{\rm v}/M_{\rm v}, \tag{1.84}$$

where $R_{\rm v}$ and $M_{\rm v}$ are the radius and the mass of the visible universe, respectively. According to the languchange of Mach's principle, (1.84) implies that the constant G is determined by the way mass is distributed in the universe. The universe evolves, and so does G. This provided another argument for a time-dependent G, and hence for a scalar–tensor theory.

It should be pointed out, however, that the above interpretation of (1.84) is by no means unique. In fact one may convince oneself that (1.84) is simply another expression of the *critical density* $\rho_{\rm cr}$ of the universe:

$$3H^2 = 8\pi G\rho_{\rm cr}, \tag{1.85}$$

where the Hubble parameter H is defined by

$$H = \dot{a}/a, \tag{1.86}$$

with the scale factor $a(t)$, which varies as a power of t with the exponent not very much different from unity, also assuming that $R_{\rm v} \sim t$. According to the standard interpretation, we assume a constant G, and ask whether the observed Hubble parameter and the density satisfy (1.85). In this context one has to choose the nature of three-dimensional space; closed, flat, or open.

We in fact find a much smaller density than the critical density, and ascribe the deficit to "dark matter," and possibly also to "dark energy." As will be shown in Chapter 5, this was precisely what triggered the emergence of the problem of the cosmological constant in the contemporary sense. In this way, the relation which once, under the name of Mach's principle, served as a motivation for creating the scalar–tensor theory now re-emerges to provide a new springboard in modern cosmology.

1.3.5 Does the scalar–tensor theory have any advantage over simple scalar theories?

We see many examples of theories or models of a gravitational scalar field, which are less restricted than the one in the scalar–tensor theory. Most relevant at this time is an approach that goes by the name of quintessence. Does one gain any advantage by sticking to the complicated theory?

We first re-iterate that many of the theoretical models of unification feature the nonminimal coupling term. Also to be noticed is the fact that any theory including a scalar field can be made expressible in terms of a nonminimal coupling term, as will be emphasized in Chapter 3. Secondly, a theoretical constraint usually implies the presence of interrelations among results that would have been left separated otherwise. We know the excellent example of gauge theories. One might gain a reward by working under tighter conditions.

As an example taken from discussions in this book, we emphasize that violation of the WEP should follow from a revision as a remedy of the cosmological aspects, thus providing a natural motivation of non-Newtonian gravity, though without any solid evidence so far. A similar example is a chain relation starting from the positivity requirement on the energy density of matter, proceeding to a lower bound on ζ, a parameter which determines the strength of the scalar-field–matter coupling, and eventually the coupling strength of the non-Newtonian force.

If one of the results thus correlated to each other turns out to be excluded experimentally, the whole theoretical structure should be affected. This might be contrasted to a less constrained theoretical framework in which remedies can be offered rather arbitrarily, in general.

Toward the very end of this book we provide yet another example of this type of interconnection, by presenting a still preliminary analysis of a possible interplay between time-variability of the fine-structure constant on the one hand, and the accelerating universe on the other. We hope that further observational studies on the time-variation of coupling constants will reveal more about the scalar–tensor theory, and eventually about unified theories.

2

The prototype Brans–Dicke model

As was emphasized in the preceding chapter, the way the scalar field enters the arena of the scalar–tensor theory is not simple. It does so through what is known as a nonminimal coupling term. This is a unique feature shared by those models qualified to be called scalar–tensor theories in the sense conceived by Jordan. In spite of the simplicity of wanting to implement a variable gravitational "constant," this term is a somewhat contrived technical device that tends to obscure other issues of physical significance. One of the emphases in this chapter is placed on revealing them beyond mathematical manipulations.

Among several versions, or models, of the scalar–tensor theory, the one due to Brans and Dicke might be viewed as a "prototype." This model, which is based on certain assumptions made for the sake of simplicity, is in fact over-simplified from the point of view of theoretical models of the modern unification program. Also for some other reasons, this model may not be accepted as fully realistic. Nevertheless, a prototype has its own merit that deserves careful and comprehensive understanding. In this chapter we introduce the original BD model as a basis of the subsequent developments.

Section 1 is an elementary but technical introduction to the prototype BD model as a basis of the whole discussion that follows. In section 2, we develop the weak-field approximation which is useful in many applications. The parameterized post-Newtonian approximation, which has provided the most accurate determination of the parameter ω, is explained in section 4, though one of our later conclusions recommends that one should depart from this approach. Fermions are obviously important from a practical point of view, though a little complication is unavoidable, as is sketched in section 5. Section 6 is devoted to an intriguing discussion on the field mixing, as well as the role played by the nonminimal coupling term.

2.1 The Lagrangian

We start with the fundamental Lagrangian (1.12) re-expressed in our own notation:

$$\mathcal{L} = \sqrt{-g} \left(\tfrac{1}{2}\xi\phi^2 R - \tfrac{1}{2}\epsilon g^{\mu\nu} \, \partial_\mu\phi \, \partial_\nu\phi + L_{\text{matter}} \right). \tag{2.1}$$

The original authors used the notation ϕ for the scalar field, but we here changed it to φ defined by (1.9):

$$\varphi = \tfrac{1}{2}\xi\phi^2, \tag{2.2}$$

with the original symbol ω re-expressed in terms of ξ defined by (1.10):

$$\epsilon\xi^{-1} = 4\omega, \tag{2.3}$$

where $\epsilon = \text{Sign}\,\omega$, so that ξ is always positive. We point out again that the matter Lagrangian L_{matter} is assumed not to contain ϕ. As yet another assumption, there is no mass term for the scalar field.

In the cosmological setting, as will be discussed later, φ is a function of the cosmic time to a first approximation, allowing us to consider the effective gravitational constant depending on time. This was in fact one of the motivations which led Jordan and his followers to investigate their models. We emphasize, however, that the rate of the time-variability, if there is any, should be very slow, of the order of 10^{-10} or less in a year. Insofar as physical phenomena in the ordinary sense are concerned, therefore, the gravitational constant can be taken practically as being truly constant. The possible time-dependence can be significant only when we consider it on the cosmological time scale.

It might be useful to explain why the above term is called a "nonminimal" coupling. According to Einstein, the equivalence principle (EP) has as its ultimate expression that tangent space attached to any world point of curved space-time should be Minkowskian. Let us call this the ultimate equivalence principle (UEP), as a theoretical outgrowth of what is now called the weak equivalence principle (WEP), stating that any object under the influence only of the gravitational force falls locally with a common acceleration. The UEP provides a powerful recipe from which a physical law in the presence of gravity is obtained from the one that applies in its absence. An example for a scalar field ϕ is given.

We start by defining a theory in flat Minkowski space-time, then we apply the substitution rule

$$\eta_{\mu\nu} \to g_{\mu\nu} \quad \text{and} \quad \partial_\mu \to \nabla_\mu. \tag{2.4}$$

This is the rule called a "comma-to-semicolon rule" [42], as stated in the preceding chapter, because a simple differentiation $\partial_\mu F$ for a tensorial field $F(x)$ is often expressed by $F_{,\mu}$, while a covariant derivative $\nabla_\mu F$ is expressed by $F_{;\mu}$, though we have an exception for a scalar field; $f_{;\mu} = f_{,\mu}$.

According to this standard method, the second term on the right-hand side of (2.1) is obtained from

$$-\tfrac{1}{2}\epsilon\eta^{\mu\nu}\,\partial_\mu\phi\,\partial_\nu\phi.$$

In this context, insofar as the second term of (2.1) is concerned, the field ϕ comes to couple to gravity only through $\sqrt{-g}g^{\mu\nu}$. The gravitational coupling obtained by applying this "minimum" rule is called a minimal coupling. The first term on the right-hand side of (2.1) cannot be obtained by this rule; in flat Minkowski space-time this term simply goes away. This is the origin of the name "nonminimal."

Why did BD choose a "nonstandard" definition of φ? As will be shown shortly, we find some situations in which φ, the whole factor that multiplies $\sqrt{-g}R$, plays a more fundamental role than ϕ. On the other hand, however, some intuitive interpretation may not apply unless the kinetic term is canonical. Risks of this kind can be avoided if we use ϕ.

A scalar field with its kinetic term in the canonical form has the dimension of mass, which provides a useful index when we evaluate divergences in relativistic quantum field theory. As we can find from (2.2), however, φ has the dimension of mass squared, which is again a source of potential confusion.

Notice, however, that (1.9) implies the restriction that φ is only positive-valued. In fact we chose $\xi > 0$ in (1.9), thus imposing the restriction that $\varphi > 0$. Formulation in terms of ϕ appears to be too restricted. This is, however, an advantage if the theory is to be realistic, because negative φ implies negative G_{eff}, hence "antigravity." It is certainly unnatural if dynamical evolution of the scalar field would lead to this undesirable situation of going through an infinitely strong gravitational effect.

We point out another important point in the Lagrangian (2.1). It is assumed that the matter Lagrangian L_{matter} does not contain ϕ. This is necessary in order to ensure that the WEP applies, as we alluded in Chapter 1. One may wonder whether this assumed decoupling from matter implies the total absence of the coupling to matter of the scalar field, hence almost losing detectability of the scalar field. We will show that the scalar field does couple to matter *in the field equation, though not in the Lagrangian*. The decoupling in the Lagrangian imposes a restriction on how the field couples to matter. This is in fact what the WEP implies. This will be shown more rigorously, but will be understood in a simpler manner as follows.

Consider a point mass. In general relativity, its equation of motion is given by a *geodesic*, which is derived from an action

$$I_m = -m \int d\tau, \tag{2.5}$$

where τ is a proper time. The inertial mass m appears only as an overall coefficient, so it does not affect the trajectory in space-time. This implies *universal free-fall*, a major expression of the WEP.

If the scalar field has a coupling to matter at the level of the Lagrangian, it will enter (2.5) through the presence of the "source" on the right-hand side. Generally speaking, the equation of motion will contain m beyond the extent to which it is simply factored out, hence *violating* the WEP. It will then follow that, even in the absence of "other" interactions, such as the electromagnetic and the strong interactions, the energy of a point mass need not be conserved covariantly. Dicke considered that this is unacceptable in view of the results of high-precision experiments, including his own improvement of the Eötvös experiment [39], on the decoupling from matter of the scalar field. We will also show that the situation looks the same if matter is fields, rather than point masses.

2.2 Field equations

We derive field equations from (2.1). Since this involves pretty complicated manipulations, we show the results first:

$$2\varphi G_{\mu\nu} = T_{\mu\nu} + T^{\phi}_{\mu\nu} - 2(g_{\mu\nu} \Box - \nabla_\mu \nabla_\nu)\varphi, \tag{2.6}$$

$$\Box\varphi = \zeta^2 T, \tag{2.7}$$

$$\nabla_\mu T^{\mu\nu} = 0. \tag{2.8}$$

Here we use φ as defined by (2.2), although we chose ϕ as an independent field. Of course, (2.6) is derived by varying (2.1) with respect to $g^{\mu\nu}$ (multiplied by the factor 2 in accordance with the factor $\frac{1}{2}$ in (2.9) below defining the energy–momentum tensor). The emergence of 2φ may be easily understood, but the details of deriving terms involving the second derivatives of φ are given in Appendix C.

The energy–momentum tensor of matter, $T_{\mu\nu}$, is defined by

$$\frac{\delta(\sqrt{-g}L_{\text{matter}})}{\delta g^{\mu\nu}} = -\frac{1}{2}\sqrt{-g}T_{(\mu\nu)}, \tag{2.9}$$

where the notation $(\mu\nu)$ implies symmetrization;

$$A_{(\mu\nu)} \equiv \frac{1}{2}(A_{\mu\nu} + A_{\nu\mu}).$$

The energy–momentum tensor of a fermion field is not completely symmetric in general. We must use the *tetrad*, which is more fundamental than the metric. Without going into the details, for the time being, we may simply take it to be symmetric. Also T is the trace of $T_{\mu\nu}$;

$$T = g^{\mu\nu}T_{\mu\nu}. \tag{2.10}$$

On the other hand, $T^{\phi}_{\mu\nu}$ is the energy–momentum tensor of ϕ defined in the same way as in (2.9) with L_{matter} replaced by

$$L^{\phi} = -\tfrac{1}{2}\epsilon g^{\mu\nu}\,\partial_{\mu}\phi\,\partial_{\nu}\phi. \tag{2.11}$$

An explicit calculation gives

$$T^{\phi}_{\mu\nu} = \epsilon\Big[\partial_{\mu}\phi\,\partial_{\nu}\phi - \tfrac{1}{2}g_{\mu\nu}(\partial\phi)^2\Big], \tag{2.12}$$

where we used a simplified notation

$$(\partial\phi)^2 \equiv g^{\mu\nu}\,\partial_{\mu}\phi\,\partial_{\nu}\phi.$$

Equation (2.7) can be obtained in the following manner. First we vary (2.1) with respect to ϕ, obtaining

$$\xi\phi R + \epsilon\,\Box\phi = 0, \tag{2.13}$$

where $\Box$ is a covariant D'Lambertian for a scalar field, defined by

$$\Box\phi = \frac{1}{\sqrt{-g}}\,\partial_{\mu}\big(\sqrt{-g}g^{\mu\nu}\,\partial_{\nu}\phi\big). \tag{2.14}$$

Multiplying (2.13) by ϕ yields

$$2\varphi R + \epsilon\phi\,\Box\phi = 0. \tag{2.15}$$

On the other hand, we take the trace of (2.6). By using

$$g^{\mu\nu}G_{\mu\nu} = -R, \tag{2.16}$$

we obtain

$$-2\varphi R = T - \epsilon(\partial\phi)^2 - 6\,\Box\varphi. \tag{2.17}$$

We eliminate φR from (2.15) and (2.17), obtaining

$$\epsilon\Big(\phi\,\Box\phi + (\partial\phi)^2\Big) + 6\,\Box\varphi = T. \tag{2.18}$$

By utilizing the relation

$$\Box\phi^2 = 2\Big(\phi\,\Box\phi + (\partial\phi)^2\Big), \tag{2.19}$$

we finally obtain (2.7)

$$\Box\varphi = \zeta^2 T, \tag{2.20}$$

where we introduced ζ from (1.14):

$$\zeta^{-2} = 6 + \epsilon\xi^{-1} = 6 + 4\omega. \tag{2.21}$$

Equation (2.20) shows that the scalar field has its source given by the trace of the matter energy–momentum tensor. As mentioned before, the scalar field is decoupled from matter in the Lagrangian. How does it couple to matter in the field equation? We will discuss this question later. Notice also that the simple form (2.20) emerges only for φ rather than ϕ, which is an advantage of using φ, as mentioned before.

According to (2.20) and (2.21), the coupling to matter of the scalar field vanishes in the limit $\xi \to 0$, or $\omega \to \infty$. This agrees with the limit $\zeta \to 0$, as one finds by using (2.21). In other words, the theory tends to Einstein's theory in this limit. This is the crucial meaning of the parameter ξ or ω. Even in this limit, the scalar field would play the role of the cosmological *dark matter*, as will be discussed in later chapters.

From a dimensional analysis, we find that ζ has the dimension of mass^{-2} (notice that φ has the dimension of mass^2), which is supplied by $M_P^{-2} = 8\pi G$. More explicitly, we may write

$$\zeta^2 = M_P^{-2}\frac{1}{6 + \epsilon\xi^{-1}} = \frac{4\pi G}{3 + 2\omega}. \tag{2.22}$$

We now move on to derive (2.8). In principle it can be derived from equations of motion of matter. As we are going to show, however, we can derive it directly from the assumed absence of the coupling to matter of ϕ, independently of specific properties of individual matter fields. This is in fact a consequence of application of the Bianchi identity to (2.6).

Put (2.6) into the form of a contravariant vector, and apply ∇_μ to the result to obtain

$$\nabla_\mu T^{\mu\nu} = -\nabla_\mu T^{\mu\nu}_\phi + 2([\nabla^\nu, \Box] + G^{\mu\nu}\,\partial_\mu)\varphi, \tag{2.23}$$

where we have used the Bianchi identity $\nabla_\mu G^{\mu\nu} = 0$. The last term on the right-hand side is a contribution from the part in which ∇_μ operates on φ. It is possible to show that the right-hand side of (2.23) vanishes identically, though we must go through considerable complications, which are given in Appendix D.

In this way (2.8) is derived, showing that the covariant conservation law of matter remains unaffected by the presence of the scalar field. If a term for the ϕ–matter coupling were present in (2.1), the right-hand side of (2.13) would have acquired a corresponding contribution, and so would the right-hand side of (2.23). Then (2.8) would have failed to hold. Readers are advised to refer also to Appendix D on this detail.

If we consider a point mass as matter, this covariant conservation law provides a necessary and sufficient condition for the world line to be a geodesic. Also, the world line's being a geodesic assures that the WEP holds. It thus follows that the WEP is assured to hold by the absence of the scalar field in L_{matter}.

2.3 The weak-field approximation

We try to see what physical results can be derived from these field equations. We first study the limit in which the fields are weak. This would correspond to the Newtonian approximation of Einstein's equation.

We write for the scalar field

$$\phi(x) = v + Z\sigma(x), \tag{2.24}$$

where v is a constant with the dimension of mass, corresponding to a "vacuum expectation value" in the quantum field theory. Also it will play the role of the "background field" in cosmology, as will be discussed later. Z is another constant that will be determined later. We assume that

$$v \gg Z\sigma, \tag{2.25}$$

to linearize the equations with respect to σ.

We have $\phi = v$ to the zeroth approximation for which the first term on the right-hand side of (2.1) is simply

$$\sqrt{-g}\frac{1}{2}\xi v^2 R. \tag{2.26}$$

Since this has the same appearance as the Einstein–Hilbert term (1.7), we reach the identification

$$\xi v^2 = M_{\text{P}}^2 = 1. \tag{2.27}$$

We learn that v is essentially the Planck mass. More precisely, in the reduced Planckian system, however, we should have

$$v = \xi^{-1/2}, \tag{2.28}$$

from (2.27).

From (2.24) and (2.28) we now have

$$\varphi \approx \tfrac{1}{2} + \xi^{1/2} Z\sigma. \tag{2.29}$$

On substituting this into (2.7) we obtain

$$\Box \sigma = \xi^{-1/2} Z^{-1} \zeta^2 T. \tag{2.30}$$

The coefficient on the right-hand side will eventually be fixed to ζ, after the factor Z is determined to be $\xi^{-1/2}\zeta$, either through the consistency condition for a special example (see (2.118)), or more generally in the formulation of conformal transformation. For the time being, however, the factor is left undetermined, still allowing us to reach a useful result.

In the Newtonian limit, we consider matter objects at rest. In terms of the mass density ρ, we write

$$T = -\rho. \tag{2.31}$$

Since obviously the solution σ of (2.30) is time-independent, (2.30) will be reduced to

$$\vec{\nabla}^2 \sigma(\vec{r}) = -\xi^{-1/2} Z^{-1} \zeta^2 \rho(\vec{r}). \tag{2.32}$$

For a point mass with the total mass M, this Poisson equation has a solution

$$\sigma(\vec{r}) = \frac{M}{M_{\mathrm{P}}} \xi^{-1/2} Z^{-1} \zeta^2 \frac{1}{4\pi r}, \tag{2.33}$$

where we supplied M_{P}^{-1} on the right-hand side in order to indicate the correct dimension.

Linearization with respect to the metric field can also be applied in the same way. As in the conventional technique, we introduce the field $h_{\mu\nu}(x)$ by putting

$$g_{\mu\nu} = \eta_{\mu\nu} + M_{\mathrm{P}}^{-1} h_{\mu\nu}(x). \tag{2.34}$$

We expand $G_{\mu\nu}$ up to terms linear in $h_{\mu\nu}(x)$, obtaining

$$G_{\mu\nu} = \frac{1}{2M_{\mathrm{P}}} \Big[-\Box h_{\mu\nu} + \partial_\mu \partial_\lambda h_\nu^\lambda + \partial_\nu \partial_\lambda h_\mu^\lambda - \partial_\mu \partial_\nu h$$

$$- \eta_{\mu\nu} (\partial_\rho \partial_\sigma h^{\rho\sigma} - \Box h) \Big], \tag{2.35}$$

where

$$h = h_\mu^\mu. \tag{2.36}$$

Notice that the suffices are raised or lowered by $\eta_{\mu\nu}$. Substitute (2.29) and (2.36) into (2.6) with $\phi = v$ on the left-hand side. Dropping terms of

$\mathcal{O}(\sigma^2)$, we may put

$$T^\phi_{\mu\nu} \approx 0. \tag{2.37}$$

On the other hand, we have

$$\Box\phi^2 \approx 2vZ\,\Box\sigma, \tag{2.38}$$

$$\nabla_\mu\nabla_\nu\phi^2 \approx 2vZ\,\partial_\mu\partial_\nu\sigma, \tag{2.39}$$

which are used to put (2.6) into the form

$$\xi\frac{v^2}{2M_{\rm P}}(-\Box h_{\mu\nu} + \cdots) = T_{\mu\nu} - 2vZ\xi(\eta_{\mu\nu}\,\Box - \partial_\mu\partial_\mu)\sigma. \tag{2.40}$$

By rearranging terms and eliminating v by using (2.28), we finally put (2.6) into the form

$$\Box h_{\mu\nu} - \partial_\mu\partial_\lambda h^\lambda_\nu - \partial_\nu\partial_\lambda h^\lambda_\mu + \partial_\mu\partial_\nu h + \eta_{\mu\nu}(\partial_\rho\partial_\sigma h^{\rho\sigma} - \Box h)$$
$$-4Z\xi^{1/2}(\eta_{\mu\nu}\,\Box - \partial_\mu\partial_\mu)\sigma = -2M_{\rm P}^{-1}T_{\mu\nu}. \tag{2.41}$$

The last two terms on the left-hand side represent *mixing* between two fields, $h_{\mu\nu}$ and σ. Remember that they come from the nonminimal coupling term.

We may remove this mixing by applying the process of *diagonalization*. For this purpose we introduce another field $\chi_{\mu\nu}$ defined by

$$\chi_{\mu\nu} = h_{\mu\nu} - \tfrac{1}{2}\eta_{\mu\nu}h - 2Z\xi^{1/2}\eta_{\mu\nu}\sigma. \tag{2.42}$$

This is inverted to give

$$h_{\mu\nu} = \chi_{\mu\nu} - \tfrac{1}{2}\eta_{\mu\nu}\chi - 2Z\xi^{1/2}\eta_{\mu\nu}\sigma, \tag{2.43}$$

which is then substituted into (2.40). We also impose a coordinate condition

$$\partial_\lambda\chi^\lambda_\nu = 0, \tag{2.44}$$

to obtain the equation without mixing

$$\Box\chi_{\mu\nu} = -\frac{2}{M_{\rm P}}T_{\mu\nu}. \tag{2.45}$$

For a point mass at rest with mass M, the solution of (2.45) is given by

$$\chi_{00}(\vec{r}) = 2\frac{M}{M_P}\frac{1}{4\pi r}, \qquad \text{other components} = 0. \tag{2.46}$$

Let us consider another point mass with mass m, which moves in the fields $\chi_{\mu\nu}$ and σ. As stated before, the equation of motion of a point mass is a geodesic. This implies that the prototype BD model falls into the category of "metric theories," in which equations of motion of matter are determined by the space-time metric. The latter is affected, however, by the matter in a manner different from the way in which it is affected in pure general relativity. In this sense, the "gravitational" potential is still given by

$$V = -\frac{1}{2}\frac{m}{M_P}h_{00}. \tag{2.47}$$

According to (2.43) we find

$$h_{00} = \chi_{00} + \tfrac{1}{2}\chi + 2Z\xi^{1/2}\sigma = \tfrac{1}{2}\chi_{00} + 2Z\xi^{1/2}\sigma, \tag{2.48}$$

where we used $\chi = -\chi_{00}$. In this way we obtain

$$V = V_\chi + V_\sigma, \tag{2.49}$$

where

$$V_\chi = -\frac{1}{4}\frac{m}{M_P}\chi_{00}, \tag{2.50}$$

$$V_\sigma = -\frac{m}{M_P}Z\xi^{1/2}\sigma. \tag{2.51}$$

On further substituting from (2.46) and (2.33) we obtain

$$V_\chi = -\frac{1}{2}\frac{mM}{M_P^2}\frac{1}{4\pi r}, \tag{2.52}$$

$$V_\sigma = -\frac{mM}{M_P^2}\zeta^2\frac{1}{4\pi r}. \tag{2.53}$$

As we alluded before, the result for V_σ is independent of the coefficient Z.

We notice that the portion V_χ from the tensor field and V_σ from the scalar field share exactly the same dependences on the masses and the distance. This implies that the force due to the scalar field couples to matter in proportion to the inertial mass of an object. It follows that any object falls with the common acceleration even with the contribution of the scalar force included. The WEP continues to hold true. This is obviously due to the fact that the source of the scalar field is T, the trace of the matter energy–momentum tensor. We can further trace its origin back to the situation that the scalar field couples to $T_{\mu\nu}$ only through the mixing interaction. The component T_{00}, for example, is the total energy of the system, independent of what portion comes from the kinetic energies of constituent particles and what part from the potential energies, and so on. In this sense the WEP is often called *composition-independence*, which general relativity was designed to derive automatically. This outstanding feature is inherited also by the scalar field in the prototype BD model.

Combining (2.52) and (2.53) yields

$$V = -G_1 m^2 / r, \tag{2.54}$$

where

$$G_1 = \frac{4 + 2\omega}{3 + 2\omega} G. \tag{2.55}$$

We started out by considering that the coefficient G in the Einstein–Hilbert term (1.7) is the fundamental gravitational constant. However, it is G_1 that is measured by experiments. G is the gravitational constant only for the tensor gravitational force. Newton's constant measured in any physical situation is G_1 that includes the contribution from the scalar force as well. There is no way to separate these two contributions as long as we measure the force acting between static objects. We can separate them, however, if we go to the level of general relativity, which will be the subject of the next section. We can do so also if the scalar field is finite-ranged, as is expected for non-Newtonian gravity.

According to (2.55), we find $G_1 \to G$ in the limit $\omega \to \infty$. This is because the scalar coupling to matter vanishes in this limit.

We also note from (2.53) that the force due to the exchange of σ is attractive as long as

$$\zeta^2 > 0 \tag{2.56}$$

is obeyed. This is another expression of the fact that σ is a normal field, instead of a ghost. According to (2.21), this can be true even if $\epsilon = -1$, and

$$-\tfrac{3}{2} < \omega < 0, \quad \text{or} \quad \xi > \tfrac{1}{6}. \tag{2.57}$$

This indicates, as will be discussed in Chapter 3, that the "physical" σ need not necessarily be a ghost even if the original field ϕ is.

2.4 The parameterized post-Newtonian approximation

We are going to see how the prototype BD theory fits the observations. For this purpose, we go another step further from the Newtonian approximation to the *parameterized post-Newtonian* (PPN) *approximation*. We start by expanding the Schwarzschild metric into power series of Schwarzschild radius/distance up to the first and the second order for $g_{rr}(r)$ and $g_{00}(r)$, respectively. We then extend the result by allowing the coefficients of the expansion to be arbitrary for a class of theories not exactly the same as general relativity. In terms of what are called Eddington's parameters β and γ, we write

$$-g_{00} \approx 1 - \frac{a_g}{r} + \frac{\beta - \gamma}{2} \frac{a_g^2}{r^2}, \tag{2.58}$$

$$g_{rr} \approx 1 + \gamma \frac{a_g}{r}, \tag{2.59}$$

where a_g is the Schwarzschild radius. Details of this formulation are summarized in Appendix E.

The original Schwarzschild solution corresponds to the special choice

$$\beta = \gamma = 1. \tag{2.60}$$

General relativity can be tested by studying how accurately (2.60) is verified. Solar-system experiments, for example, have been analyzed in terms of these parameters. Among well-known examples, we know about the

- deflection of light around the Sun:

$$\Delta\varphi = \frac{1+\gamma}{2}\left(\frac{2a_g}{r_0}\right),$$

- perihelion shift of Mercury:

$$\Delta\varphi = \frac{2 - \beta + 2\gamma}{3}\left(\frac{3a_g\pi}{\ell}\right),$$

- and Shapiro delay (radar echo):

$$\Delta t = 2a_g\left[1 + \frac{1+\gamma}{2}\ln\left(\frac{4r_1r_2}{r_0^2}\right)\right].$$

In the Viking Project [6], the most accurate result for γ was obtained from the measurement of the Shapiro delay; $\gamma = 1 \pm 0.001$. The latest result for deflection of light, from very-long-baseline-interferometry (VLBI) observations of radio waves from extragalactic radio sources, yielded [7]

$$\gamma = 1.000\,00 \pm 0.000\,28. \tag{2.61}$$

Since the prototype BD theory is a metric theory, motion of matter objects can be described in the framework of the PPN approximation. We are going to show that the major effect of the scalar field occurs in the coefficient of the $1/r$ term of g_{rr}, namely through shifting γ from unity. To this lowest-order approximation, we can ignore the $1/r^2$ term of g_{00}. The final result is (2.84), and the constraint on the parameter ω is given by (2.87), though they will be obtained by somewhat lengthy calculations.

Let us start with the "modified version of Einstein's equation" (2.6). It is convenient to transform it such that only $R_{\mu\nu}$ will appear on the left-hand side by using (2.17). We also choose

$$T_{\mu\nu} = 0, \tag{2.62}$$

since we are here interested in the "external solution."

To this approximation we may solve (2.7) by assuming the same type of solution as (2.29) with (2.33);

$$\varphi \approx \frac{1}{2} + \zeta^2 \frac{m}{4\pi r}. \tag{2.63}$$

By substituting this into (2.12) we find

$$T^{\phi}_{\mu\nu} \sim \mathcal{O}(r^{-4}), \tag{2.64}$$

which may be ignored. Notice that no time-derivatives appear since we are looking at a static situation. After these considerations we now have (2.6) in the form

$$2\varphi R_{\mu\nu} = (g_{\mu\nu} \Box + 2\nabla_{\mu}\nabla_{\nu})\varphi \equiv K_{\mu\nu}. \tag{2.65}$$

On the left-hand side, we follow the usual prescription;

$$R_{00} = e^{-\lambda+\nu} \left(\frac{1}{2}\nu'' + \frac{1}{4}\nu'^2 - \frac{1}{4}\nu'\lambda' + \frac{\nu'}{r} \right), \tag{2.66}$$

$$R_{rr} = -\frac{1}{2}\nu'' - \frac{1}{4}\nu'^2 + \frac{1}{4}\nu'\lambda' + \frac{\lambda'}{r}, \tag{2.67}$$

where

$$g_{00} = -e^{\nu(r)}, \qquad g_{rr} = e^{\lambda(r)}. \tag{2.68}$$

$\Box\varphi$ in $K_{\mu\nu}$ can be given by (2.20). Also, from (2.62), we obtain

$$\Box\varphi = 0. \tag{2.69}$$

From time-independence of φ we find $K_{00} = 0$, and hence

$$\frac{1}{2}\nu'' + \frac{1}{4}\nu'^2 - \frac{1}{4}\nu'\lambda' + \frac{\nu'}{r} = 0, \tag{2.70}$$

in agreement with the result in general relativity.

From (2.58) and (2.68) we may write

$$\nu = \ln(-g_{00}) \approx \ln\left(1 - \frac{a_g}{r} + \frac{\beta - \gamma}{2}\frac{a_g^2}{r^2}\right)$$

$$\approx -\frac{a_g}{r} + \frac{\beta - \gamma - 1}{2}\frac{a_g^2}{r^2}. \tag{2.71}$$

In the same way we obtain

$$\lambda = \ln(g_{rr}) \approx \ln\left(1 + \gamma\frac{a_g}{r}\right)$$

$$\approx \gamma\frac{a_g}{r}. \tag{2.72}$$

Differentiating these yields also

$$\nu' \approx \frac{a_g}{r^2} - (\beta - \gamma - 1)\frac{a_g^2}{r^3}, \tag{2.73}$$

$$\nu'' \approx -2\frac{a_g}{r^3} + 3(\beta - \gamma - 1)\frac{a_g^2}{r^4}, \tag{2.74}$$

$$\lambda' \approx -\gamma\frac{a_g}{r^2}. \tag{2.75}$$

Use these in (2.70). We retain terms only to the order a_g/r^3 and a_g^2/r^4. The terms of a_g/r^3 cancel each other out, leaving

$$\frac{-1 - \gamma + 2\beta}{4}\frac{a_g^2}{r^4},$$

which must vanish, thus leading to

$$\beta = \frac{1 + \gamma}{2}. \tag{2.76}$$

We now move on to R_{rr}. By combining it with $R_{00} = 0$, we obtain

$$R_{rr} = \frac{\nu' + \lambda'}{r}. \tag{2.77}$$

By using (2.75) we find

$$R_{rr} \approx (1 - \gamma)\frac{a_g}{r^3}, \tag{2.78}$$

where we dropped a_g^2/r^4.

We next compute the second term of K_{rr};

$$\nabla_r \nabla_r \varphi = (\partial_r - \Gamma^r_{rr})\varphi'(r), \tag{2.79}$$

$$\begin{aligned} \Gamma^r_{rr} &= \tfrac{1}{2}g^{rr}\,\partial_r g_{rr} = \tfrac{1}{2}\lambda'(r) \\ &= -\frac{\gamma}{2}\frac{a_g}{r^2}. \end{aligned} \tag{2.80}$$

We have used (2.75) in the last step. By substituting this into (2.79), and using (2.63), we obtain

$$K_{rr} \approx \zeta^2 \frac{m}{\pi}\frac{1}{r^3}, \tag{2.81}$$

where we have again dropped the term a_g/r^4. By comparing this with (2.78), we find that they share the same $1/r^3$ form, thus obtaining

$$1 - \gamma = \zeta^2 \frac{m}{\pi a_g}. \tag{2.82}$$

Since the Schwarzschild radius is a gravitational effect due to the tensor field, it must be given in terms of G;

$$a_g = 2Gm = \frac{2m}{8\pi} = \frac{m}{4\pi}, \tag{2.83}$$

from which follows finally

$$\gamma = 1 - 4\zeta^2. \tag{2.84}$$

In this way we established that ζ^2 represents the deviation from pure general relativity. We have confirmed that $\gamma \to 1$ for $\omega \to \infty$.

We also have

$$\beta = 1 - 2\zeta^2, \tag{2.85}$$

from (2.76).

According to (2.61), we may allow the prototype BD model only if the strength of coupling to matter is limited by

$$4\zeta^2 \lesssim 2.8 \times 10^{-4}, \tag{2.86}$$

or, by using (1.10) and (1.14),

$$\omega \gtrsim 3.6 \times 10^3, \quad \text{or} \quad \xi \lesssim 7.0 \times 10^{-5}. \tag{2.87}$$

This is the most stringent constraint available at present, since β is constrained less stringently by an order of magnitude.

There was an interesting episode about the "solar oblateness." Dicke must have expected ω to be naturally as large as of the order of unity, instead of such a large value as indicated by (2.87). How much would general relativity be affected in that case? A planet, Mercury, for example, will exhibit a perihelion shift during a revolution of

$$\Delta\varphi = \frac{2 - \beta + 2\gamma}{3}\left(\frac{3a_g\pi}{\ell}\right) \tag{2.88}$$

according to general relativity. Using the solar Schwarzschild radius for a_g, and the average radius of Mercury's orbit for ℓ, we find

$$\Delta\varphi = \frac{2 - \beta + 2\gamma}{3} \times 43.03'' \text{ century}^{-1}. \tag{2.89}$$

The value 43 arcseconds beautifully filled the gap between calculation due to Newtonian mechanics and observation, hence giving Einstein confidence for the first time that his theory was correct. This shows that $(2 - \beta + 2\gamma)/3$ is very close to 1.

According to (2.21), (2.84), and (2.85), on the other hand, we should have

$$\frac{2 - \beta + 2\gamma}{3} = 1 - 2\zeta^2 = \frac{\epsilon\omega + 1}{\epsilon\omega + \frac{3}{2}}. \tag{2.90}$$

Suppose that $\omega = 1$, for example. We would find

$$\text{right-hand side of (2.90)} = \begin{cases} 0.8, & \text{if } \epsilon = +1, \\ 0, & \text{if } \epsilon = -1. \end{cases}$$

This would certainly jeopardize the crucial result of 43 arcseconds per century.

Dicke then suspected, as was mentioned in Chapter 1, that there was a flaw in comparing theory and observation. More specifically, he raised the question of whether the solar mass distribution might not be fully spherically symmetric. Remember that (2.88) is a consequence of assuming the spherical symmetry of the mass distribution and hence of geometry.

He showed that the coefficient of $3a\pi/\ell$ in (2.88) should acquire a correction

$$\frac{a_\odot}{2m}J_2 \approx 3 \times 10^{-4}\left(\frac{J_2}{10^{-7}}\right), \tag{2.91}$$

if the Sun has a small *quadrupole moment* J_2, where $a_\odot$ is the average solar radius, while m is the mass of Mercury. Owing to the solar spin, of course, there is centrifugal force, generating oblateness, and hence J_2 can be estimated to be $\sim 10^{-7}$. This would give $\sim 10^{-4}$ for (2.91), far smaller than the value of 1, the result from general relativity.

Dicke made his own measurement of the shape of the Sun, concluding that $J_2 \sim 10^{-5}$ [8]. This allows $(2 - \beta + 2\gamma)/3$ to deviate from unity by

~0.1, and hence

$$\omega \approx 5, \qquad \text{or} \qquad \omega^{-1} \approx 0.2, \tag{2.92}$$

which might be viewed to be of the order of unity.

This controversial argument prompted many researchers to repeat the measurement, which, however, needs great care. All of them, including the latest measurements from the VLBI experiment [7], ended up with the result $J_2 \lesssim 10^{-6}$, leaving the question of why Dicke obtained such a large value still unanswered.

Another point should be mentioned. We have considered the scalar field to be massless. This is one of the crucial assumptions in the prototype model, which was probably made just for simplicity to start with. Since, however, no known principle forbids a scalar field to have a nonzero mass, this is likely subject to modification due to quantum effects, for example, as will be discussed in section 4. If the force-range of the scalar field turns out to be shorter than the size of the solar system, the parameters of the BD model are no longer constrained by the solar-system experiments.

2.5 A spinor field as matter

We have limited ourselves to considering matter simply as a mass distribution without asking any more details about its content. If we extend our discussion to unified theories, for example, we must take the fact that matter consists of elementary particles into account. Among them fermions pose intrinsic complications. In later chapters, we will rarely use detailed expressions for fermions except in Appendix F. Nevertheless, it is convenient to work on fermions, unlike bosons, in order to discuss a classical particle picture in the field-theoretical approach, as will be shown in this section. We review briefly how we can incorporate fermions, more specifically spinor fields, into the arena of gravity, only to the extent that we derive the relation (2.118) and hence (2.119), for the moment, in order to make the discussion self-consistent.

The relation (2.119) is derived also in Chapter 3, (3.54). In this sense, one may skip this section unless one is particularly interested in technical aspects of fermions.

Take the simplest example of a free massive spinor field ψ. The Lagrangian in flat Minkowski space-time is given by

$$L_{\text{matter}} = -\overline{\psi}(\not{\partial} + m)\psi, \tag{2.93}$$

where m is mass, and

$$\not{\partial} = \gamma^\mu \, \partial_\mu. \tag{2.94}$$

In order to be able to incorporate a half-integer-spin field into curved space-time, we need the concept of the *tetrad*, which is more fundamental than the metric [54–56].

This is related to the fact that the concept of a spinor is unique to Minkowski space-time; we are not allowed to extend it to curved space-time. We then define ψ on flat Minkowski space-time as tangential space attached to each world point on Riemannian space-time. Two spacetimes, Minkowski and Riemannian, are connected by the tetrad field b^i_μ, in which i is an index associated with Lorentz transformation on Minkowski space-time, while μ, also running from 0 to 3, is a subscript associated with the Riemannian manifold. The tetrad is related to the metric in curved space-time by the relation

$$g_{\mu\nu} = \eta_{ij} b^i_\mu b^j_\nu = b^i_\mu b_{\nu i}. \tag{2.95}$$

We may consider b^i_μ to be a "square root" of $g_{\mu\nu}$ in the same sense as that in which the ψ is a square root of a tensor. The last equation of (2.95) shows that the index i is raised or lowered by the Minkowski metric

$$\eta_{ij} = \eta^{ij} = \mathrm{diag}(-1, +1, +1, +1). \tag{2.96}$$

We define the *inverse tetrad* $b^{\mu i}$ by

$$b^{\mu i} = g^{\mu\nu} b^i_\nu, \tag{2.97}$$

which can also be interpreted as a square root of the inverse metric:

$$g^{\mu\nu} = \eta_{ij} b^{\mu i} b^{\nu j}. \tag{2.98}$$

Like a spinor ψ, the γ matrix is defined only on Minkowski space-time. It has an index i. On the other hand, we have μ in ∂_μ in (2.94). These two different kinds of indices are connected to each other also with the help of the tetrad. The differential operator in (2.94) is now replaced by

$$\not\partial = b^{\mu i} \gamma_i \, \partial_\mu. \tag{2.99}$$

Further modification is needed, because ∂_μ is not covariant. It should be replaced by a *covariant derivative* D_μ given in terms of *spin connection* and *torsion*. Since, however, they are not physically important in the applications of our immediate interest, we simply replace D_μ by an ordinary derivative;

$$D_\mu \approx \partial_\mu. \tag{2.100}$$

In this sense we modify (2.93) to

$$L_{\text{matter}} = -\overline{\psi} \Big(b^{\mu i} \gamma_i \, \partial_\mu + m \Big) \psi, \tag{2.101}$$

which will be the basis of the following discussion. We substitute this into (2.1), varying the whole Lagrangian with respect to $\overline{\psi}$ to derive the Dirac equation on curved space-time;

$$(\not{D} + m)\psi = 0. \tag{2.102}$$

Even in Minkowski space-time, we encounter some complications about the energy–momentum tensor of a spinor field. However, the covariant conservation law (2.8) still holds true, and the trace is obtained simply to be

$$T = \overline{\psi} \not{D} \psi = -m\overline{\psi}\psi. \tag{2.103}$$

In (2.101), the effect of space-time curvature enters only through $b^{\mu i}$ as a minimal coupling. There is another contribution also through $\sqrt{-g}$, which multiplies the whole L_{matter}, resulting in no real effect since it is absorbed into the rescaled ψ. We find that the same linearized field $h_{\mu\nu}$, introduced before by (2.34), can be used in b^i_μ;

$$b^i_\mu \approx \eta^i_\mu + \tfrac{1}{2}h^i_\mu, \tag{2.104}$$

which is shown to reproduce (2.95). In this process, we have re-interpreted h^ν_μ with Greek indices as the one with mixed indices in an obvious manner.

On the other hand, we find

$$b^{\mu i} \approx \eta^{\mu i} - \tfrac{1}{2}h^{\mu i}, \tag{2.105}$$

because substituting this into (2.98) reproduces

$$g^{\mu\nu} \approx \eta^{\mu\nu} - h^{\mu\nu}, \tag{2.106}$$

on dropping terms of $\mathcal{O}(h^2)$.

Now substitute (2.105) into (2.101). Also replace ν by i in (2.42) to be used as $h^{\mu i}$. We then obtain

$$L_{\text{matter}} = -\overline{\psi}\left[\not{\partial} + m - \tfrac{1}{2}(\chi^{\mu i} - \tfrac{1}{2}\eta^{\mu i}\chi - 2Z\xi^{1/2}\eta^{\mu i}\sigma)\gamma_i\,\partial_\mu\right]\psi. \tag{2.107}$$

We put this into the form

$$L_{\text{matter}} = -\overline{\psi}(\not{\partial} + m + V)\psi, \tag{2.108}$$

where V may be interpreted as the interaction term due to the gravitational field. By further dividing this into the sum

$$V = V_\chi + V_\sigma, \tag{2.109}$$

we can write

$$V_\chi = -\tfrac{1}{2}(\chi^{\mu i} - \tfrac{1}{2}\eta^{\mu i}\chi)\gamma_i\,\partial_\mu, \tag{2.110}$$

$$V_\sigma = Z\frac{\xi^{1/2}}{M_P}\sigma\,\dot{\partial}. \tag{2.111}$$

Suppose that a particle described by ψ is nearly at rest. We may then put

$$\gamma_0 \approx i, \qquad \partial_0 \approx -im, \tag{2.112}$$

with the rest of the components taken as zero. In this way we obtain

$$V_\chi \approx -\frac{m}{2M_P}\left(\chi_{00} + \frac{1}{2}\chi\right), \tag{2.113}$$

in agreement with (2.50).

In (2.111), on the other hand, we may consider that the zeroth-order equation holds, giving

$$\dot{\partial} \approx -m, \tag{2.114}$$

which allows us to obtain

$$V_\sigma = -mZ\xi^{1/2}\sigma, \tag{2.115}$$

again in agreement with (2.51). We recognize that the scalar field occurred in the Lagrangian because we smuggled σ in explicitly through the diagonalization process (2.42).

Equation (2.115) implies that the matter Lagrangian (2.101) contains the interaction term

$$L_{\psi\sigma} = -\overline{\psi}V_\sigma\psi = mZ\xi^{1/2}\overline{\psi}\psi\sigma. \tag{2.116}$$

We also expect that, by varying the total Lagrangian including (2.116), we should be able to derive the field equation

$$\Box\sigma = -\frac{\partial L_{m\sigma}}{\partial\sigma} = Z\xi^{1/2}T, \tag{2.117}$$

where we used (2.103). On the other hand, the same equation was already obtained in (2.30), but with an apparently different right-hand side $Z^{-1}\xi^{-1/2}\zeta^2 T$. These are made to agree with each other if we determine Z as

$$Z = \xi^{-1/2}\zeta, \tag{2.118}$$

by which (2.30) is simplified to

$$\Box\sigma = \zeta T. \tag{2.119}$$

2.6 The mechanism of mixing

It does not appear that there is any mixing term in the initial Lagrangian (2.1). How did it appear in the field equation (2.20)? The key lies in the nonminimal coupling term. The crucial ingredients in deriving (2.20) are in fact (2.15) and (2.17), which contain the term $2\varphi R$ coming from the nonminimal coupling. We further find that the nonminimal coupling term played the role of making σ enter the equation for $h_{\mu\nu}$, hence causing mixing between $h_{\mu\nu}$ and σ. We add that basically the same argument about the role of the nonminimal coupling term was also presented by Santiago [57].

One might argue that there should be no mixing between $h_{\mu\nu}$ describing a field of spin 2 and the scalar field for spin 0. A closer look shows, however, that a second-rank tensor contains components of spin 1 and spin 0 as well, leaving the spin-2 component in the physical state. This spinless component couples to σ. This is related to the fact that σ occurs in (2.40) with the second derivatives.

Field mixing in field equations can also be expressed in terms of mixing in the Lagrangian. In fact, substituting (2.24) and (2.34) directly into the first term on the right-hand side of (2.1), and picking up terms linear in $h_{\mu\nu}$ and in σ, yields

$$L_{h\sigma} = \zeta\sigma(\partial_\mu\partial_\nu h^{\mu\nu} - \Box h), \qquad (2.120)$$

where we used (2.27) and (2.118) [58]. By varying this with respect to $h_{\mu\nu}$, we can establish that the last term of (2.40) can be obtained. Varying with respect to σ yields the term obtained by expanding the first term of (2.13).

As illustrated by Fig. 2.1, a matter field that couples originally only to the tensor field comes to couple to the scalar field via the coupling represented by (2.120). This is what our calculation so far has shown.

In computing the diagram in Fig. 2.1, we have divergence coming from $k = 0$, where k is the (vanishing) 4-momentum carried both by σ and by $h_{\mu\nu}$. The vanishing denominator k^{-2} is canceled out, however, by k^2 in the numerator coming from the derivatives in (2.120), $\zeta(k^2\eta_{\mu\nu} - k_\mu k_\nu)$. In this way we find that this diagram gives a finite contribution, in agreement with (2.119).

In this sense the presence of derivatives is crucially important, corresponding to the situation that mixing occurs in the kinetic term. The same situation can be found in simple mechanical systems, as will be explained.

Consider the Lagrangian

$$L = \tfrac{1}{2}\dot{q}_1^2 + \tfrac{1}{2}\dot{q}_2^2 + k\dot{q}_1\dot{q}_2 + q_1 J_1, \qquad (2.121)$$

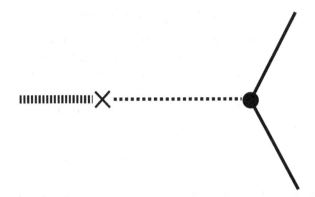

Fig. 2.1. Mixing coupling is represented by a cross between σ (thick broken line) and $h_{\mu\nu}$ (dotted line). The right-hand end, marked by a filled circle, is for the ordinary coupling to a matter field (solid line) through $T_{\mu\nu}$.

where q_i $(i = 1, 2)$ are the (generalized) coordinates of two point masses. The third term on the right-hand side represents mixing in the kinetic term, as in a double pendulum. The last term is for the effect of an external force acting on one of the masses. Another mass, subject to no direct external force, corresponds to the scalar field in the prototype BD model.

By varying (2.121), we obtain two equations of motion:

$$\ddot{q}_1 + k\ddot{q}_2 = J_1, \tag{2.122}$$
$$\ddot{q}_2 + k\ddot{q}_1 = 0. \tag{2.123}$$

The second equation can be put into the form

$$\ddot{q}_2 = -k\ddot{q}_1, \tag{2.124}$$

which can be substituted into the first equation, giving

$$\ddot{q}_1 = \frac{1}{1 - k^2} J_1, \tag{2.125}$$

obviously without mixing. By further substituting this into (2.124) we obtain

$$\ddot{q}_2 = -\frac{k}{1 - k^2} J_1, \tag{2.126}$$

showing that the external force acts finally on q_2 as well.

Equation (2.124) corresponds to (2.13), while (2.126) corresponds to (2.20), as one can easily infer. However, (2.125), as it stands, does not correspond to (2.45). Here we needed a new field $\chi_{\mu\nu}$ as a mixture of $h_{\mu\nu}$ and σ.

Owing to the lack of the potential term, we were able to solve the equations in a very simple manner. In the field-theory model, on the other hand, we must "normalize" the fields. Corresponding to this situation, we must carry out the diagonalization in the Lagrangian. The result is given by the diagonalized coordinates $\tilde{q}_i$ $(i = 1, 2)$ defined by

$$q_1 = \tilde{q}_1 - \frac{k}{\sqrt{1 - k^2}}\tilde{q}_2, \tag{2.127}$$

$$q_2 = \frac{1}{\sqrt{1 - k^2}}\tilde{q}_2, \tag{2.128}$$

where we assume that $k^2 < 1$. By substituting (2.127) and (2.128) into (2.121), we find the Lagrangian

$$L = \frac{1}{2}\dot{\tilde{q}}_1^2 + \frac{1}{2}\dot{\tilde{q}}_2^2 + \left(\tilde{q}_1 - \frac{k}{\sqrt{1 - k^2}}\tilde{q}_2\right)J, \tag{2.129}$$

in which there is no longer mixing. Equation (2.127) is the result corresponding to (2.43).

We believe that these analyses of the mechanical system serve to help the reader understand the unique features of the prototype BD model.

3

Conformal transformation

Once we introduce the nonminimal coupling term, we face the issue of *conformal transformations*. By applying a conformal transformation, we can put a nonminimal coupling term into another form. This comes from the fact that Einstein's theory is not invariant under conformal transformations. In this sense this transformation has a feature different from the gauge transformation. In the literature, however, we sometimes find confusions. We wish to provide readers with a better understanding of the issue.

We are particularly interested in how we can eliminate the factor of the scalar field in a nonminimal coupling term by transforming it into a constant. We say that we move from one *conformal frame* to another by applying a conformal transformation. The questions are then those concerning what conformal frame we live in, and on what physical grounds we are able to select which. An explicit discussion of these questions will be given in Chapter 4 on cosmological applications.

Among infinitely many conformal frames, the J(ordan) frame and the E(instein) frame are those discussed most frequently. Generally speaking, physics *looks different* in two different conformal frames. In the limit of a weak gravitational field (including a diagonalization process), however, physical conclusions remain the same.

In section 1, the concept of conformal transformation is introduced in general terms but briefly. A fact of special importance in connection with the nonminimal coupling term is discussed in section 2. Section 3 discusses the coupling to matter of the scalar field which is present in the E frame, a unique feature of the prototype BD model. In section 4, geodesic equations are revisited with a particular emphasis on the conformal transformation.

61

3.1 What is a conformal transformation?

In Chapter 2, we showed that the contribution from the scalar field yields the same force as the Newtonian force in the weak-field approximation. We discussed in detail how the scalar field which is decoupled from matter in the initial Lagrangian finally couples to matter. The key role was shown to be played by the mixing interaction between scalar and tensor gravitational fields. The argument was made in the limit of weak gravity, but there is a completely general way to arrive at the same result. This is the method of conformal transformation pointed out first by Pauli. Later Dicke discussed its physical significance [59]. The method provides a clear and powerful technique, free from mathematical ambiguity, but nevertheless requires careful consideration from the physical point of view.

A conformal transformation transforms a metric $g_{\mu\nu}$ into another metric $g_{*\mu\nu}$ according to the rule

$$g_{*\mu\nu} = \Omega^2(x) g_{\mu\nu}, \tag{3.1}$$

where $\Omega(x)$ is an arbitrary function of space-time coordinate x. This is equivalent to the transformation applied to a line element,

$$ds_*^2 = \Omega^2(x)\, ds^2. \tag{3.2}$$

One could have written the transformation function simply as $\Omega(x)$, but we use the notation $\Omega^2(x)$ because we limit ourselves to transformations that keep the sign of ds^2 unchanged.

We emphasize that this is entirely different from general coordinate transformations. On the other hand, a line element defines distance in the most fundamental sense. Therefore, (3.2) implies that it changes distance, or the standard of size, by a rate that differs from point to point on the space-time manifold. Also this is done without any direction specified. In other words, it changes the rate "isotropically." It represents expansion or contraction at the same rate in the x direction and y direction, for example. As a consequence, it leaves the angle between any two vectors invariant. This is what is meant by "conformal."

Notice that we are talking about isotropy in four-dimensional space-time. This implies that changes in spatial distance and changes in time interval should occur at the same rate.

Suppose, for the moment, that Ω is a constant independent of x. The transformation is then called a *scale transformation*. We are changing meter-sticks at a rate that is common over the entirety of space-time. Always observing special relativity, this implies that the unit of clocks is also multiplied by the same number. All physical phenomena as well as the laws behind them would remain the same under this transformation if no specific standard of meter-stick existed. Suppose that this is indeed the case. We would then say that the world is *scale-invariant*. In reality, however, there is no such invariance. An atomic clock, for example, is a device with the fundamental frequency unit provided by the Rydberg constant

$$1\,\mathrm{Ryd} = \pi \frac{m_r c Z^2 \alpha^2}{\hbar}, \tag{3.3}$$

or the latter multiplied by a certain power of the fine-structure constant

$$\alpha \equiv e^2/(4\pi\hbar c) \approx 1/137, \tag{3.4}$$

where m_r is a reduced mass defined by

$$m_r^{-1} = m_e^{-1} + M^{-1}, \tag{3.5}$$

in terms of m_e, the electron mass, and M, the mass of nucleus with atomic number Z. After all, we base our argument on the notion that masses of elementary particles are fundamental constants. Obviously, the particle masses intrinsic to each of the particles should take different values if they are measured using different meter-sticks or clocks. In this sense, the existence of masses of elementary particles breaks the scale invariance of the world.

Is this enough to make the concept of scale invariance totally meaningless? The answer is not necessarily yes. At least in phenomena at energies so high that rest masses of elementary particles can be ignored, scale invariance should provide a convenient concept from a practical point of view.

From a more fundamental point of view, on the other hand, it appears that scale invariance could be related to the question of the true origin of particle masses. In order to simplify the argument, we accept, at least for the time being, that c and $\hbar$ are true constants. It is then convenient to work in a system of units in which c and $\hbar$ are unity, thus reducing three dimensions, length, time, and mass, to a single dimension, which we might choose to be mass, though this choice is not unique. By doing so we make it clear that the inverse of mass provides units of length and time. Combining this with the finding that scale invariance would have survived if no particle masses were present makes it reasonable to suspect that particle masses and breakdown of scale invariance share the same origin.

In many situations of the modern theory of elementary particles, we start with a world in which all the particles are massless, subsequently reaching the real world in which nonzero masses arise from some kind of *spontaneous breaking* of certain symmetries. In the same sense, we might start with the scale-invariant world, perhaps at the beginning of the universe. These considerations seem to support the idea that scale invariance is of fundamental importance.

Notice, however, that scale transformation as we discussed above is a *global* transformation which is independent of coordinates. According to an idea accepted widely in modern field theories, on the other hand, every fundamental transformation in physics should be *local*. This idea has been supported very strongly ever since the successes of general relativity and gauge theories. Now localizing a scale transformation yields precisely conformal transformation. In other words, conformal transformation is a *localized scale transformation*.

Gauge theories considered to provide ultimate laws of particle physics can be traced back to Weyl's theory of 1921 [19], which turned out not to be applicable to the electromagnetic field, or any other fields now considered to be gauge fields. His theory was in fact a conformal transformation. For this reason the transformation (3.1) is often called *Weyl rescaling*. The name "gauge" has its origin in the fact that (3.2) has to do with the standard of length. General coordinate transformation is a local transformation, which Weyl attempted to generalize, arriving finally at the conformal transformation.

As we stated before, scale invariance can be realized approximately at such high energies that particle masses can be ignored. Now, at such high energies, is a conformal invariance an approximate symmetry? The answer is yes in almost any situation, but not always. Theories of massless fermions or vector fields (such as the electromagnetic field) are certainly conformally invariant. However, a massless scalar field fails to exhibit complete invariance. Conformal invariance applies to a massless fermion field only if a certain condition is satisfied. See Appendix F. Furthermore, Maxwell's theory is conformally invariant only in four dimensions.

As another example of a massless field, we know the gravitational field in general relativity, which does not possess conformal invariance. This is obvious because it contains Newton's constant G having the dimension of $(\text{mass})^{-2}$. A graviton in the limit of a weak field is in fact massless, but G serves to fix a standard of length that breaks invariance even under a global scale transformation.

This seems to have something to do with another important feature, namely that a renormalization procedure fails to apply to the metric field once it has been quantized. This has provided a motivation to nurture speculations that ultimate theories of gravitation should not contain a dimensional constant and hence should be conformally invariant. According to this view, field equations might have to satisfy differential equations of the fourth order rather than of the second order as in Einstein's theory. When we try to compare the results with reality, however, we must go to theories that would result when higher symmetries are broken, probably spontaneously supplying $8\pi G = M_{\rm P}^{-2}$. Conformal invariance should no longer hold true in realistic theories that follow in this way. Is conformal invariance totally nonsense then? The answer is again not simply yes. An example showing that it does provide a useful tool is found in theories with nonminimal coupling. Before going into detailed accounts of these theories, however, we start by studying transformation rules for certain quantities under a conformal transformation.

Suppose that we have equations expressed in terms of the metric $g_{\mu\nu}$. We try to re-express them in terms of $g_{*\mu\nu}$. We say that by doing so we move to a new *conformal frame*, just like saying that we move to a new coordinate frame by applying a coordinate transformation. This implies that we use different meter-sticks at different space-time points, though it might not necessarily be easy to have an intuitive image at this moment. We often attach a symbol $*$ to remind ourselves that we are in a new conformal frame.

In order to implement this re-expression, we first invert (3.1);

$$g_{\mu\nu} = \Omega^{-2}(x)g_{*\mu\nu}. \tag{3.6}$$

We substitute this into equations in the original conformal frame. Among the simplest, we find

$$g^{\mu\nu} = \Omega^2 g_*^{\mu\nu}, \tag{3.7}$$
$$\sqrt{-g} = \Omega^{-4}\sqrt{-g_*}. \tag{3.8}$$

We also do the same for the Christoffel symbol;

$$\Gamma^{\mu}{}_{\nu\lambda} = \tfrac{1}{2}g^{\mu\rho}(\partial_\nu g_{\rho\lambda} + \partial_\lambda g_{\rho\nu} - \partial_\rho g_{\nu\lambda}), \tag{3.9}$$

obtaining

$$\Gamma^{\mu}{}_{\nu\lambda} = \Gamma^{\mu}_{*\nu\lambda} - (f_\nu \delta^\mu_\lambda + f_\lambda \delta^\mu_\nu - f_*^\mu g_{*\nu\lambda}), \tag{3.10}$$

where $\Gamma^{\mu}_{*\nu\lambda}$ is a Christoffel symbol in the new conformal frame, which was obtained by replacing every $g_{\alpha\beta}$ in (3.9) by $g_{*\alpha\beta}$. We also used the notations

$$f \equiv \ln \Omega, \tag{3.11}$$
$$f_\nu \equiv \frac{\partial_\nu \Omega}{\Omega} = \partial_\nu f, \tag{3.12}$$

with $f_*^\mu \equiv g_*^{\mu\nu} f_\nu$. Notice in (3.12) that the suffices in the new conformal frame have been raised with the help of $g_*^{\mu\rho}$. See Appendix G for more details on deriving (3.10), though the most fundamental rules of calculation are shown by the following example:

$$\partial_\mu g_{\nu\lambda} = \partial_\mu \left(\Omega^{-2} g_{*\nu\lambda} \right)$$
$$= \Omega^{-2} \partial_\mu g_{*\nu\lambda} - 2\Omega^{-3} \partial_\mu \Omega g_{*\nu\lambda}$$
$$= \Omega^{-2} (\partial_\mu g_{*\nu\lambda} - 2f_\mu g_{*\nu\lambda}). \tag{3.13}$$

We find an overall factor Ω^{-2}, which is canceled out by Ω^2 in (3.7) when we include $g^{\mu\rho}$ to compute finally $\Gamma^\mu{}_{\nu\lambda}$.

Once we have the result on $\Gamma^\mu{}_{\nu\lambda}$, we then calculate $R^\mu{}_{\nu,\rho\sigma}$, and finally derive the transformation rule for R;

$$R = \Omega^2 (R_* + 6\,\Box_* f - 6 g_*^{\mu\nu} f_\mu f_\nu), \tag{3.14}$$

though more details of the derivation can be found again in Appendix G. We used

$$\Box_* f = \frac{1}{\sqrt{-g_*}} \partial_\mu (\sqrt{-g_*} g_*^{\mu\nu} \partial_\nu f) \tag{3.15}$$

for the D'Lambertian operator in the new conformal frame.

3.2 Nonminimal coupling

Let us go back to the starting Lagrangian (1.12), which is reproduced here:

$$\mathcal{L}_{\mathrm{BD}} = \sqrt{-g} \left(\tfrac{1}{2} F(\phi) R - \tfrac{1}{2} \epsilon g^{\mu\nu} \partial_\mu \phi \, \partial_\nu \phi + L_{\mathrm{matter}} \right), \tag{3.16}$$

where we write

$$F(\phi) = \xi \phi^2, \tag{3.17}$$

which might be generalized later.

Focus first on the first term,

$$\mathcal{L}_1 = \tfrac{1}{2} \sqrt{-g} F(\phi) R. \tag{3.18}$$

By applying a conformal transformation and using (3.8) and (3.14), we put (3.18) into the form

$$\mathcal{L}_1 = \sqrt{-g_*} \frac{1}{2} F(\phi) \Omega^{-2} (R_* + 6\,\Box_* f - 6 g_*^{\mu\nu} f_\mu f_\nu). \tag{3.19}$$

Notice the occurrence of Ω^{-2}. Since this is still arbitrary and undetermined, we choose it such that

$$F(\phi) \Omega^{-2} = 1, \tag{3.20}$$

allowing the first term of (3.19) containing R_* to become precisely the Einstein–Hilbert term. We thus have

$$\Omega = F^{1/2}. \tag{3.21}$$

The second term of (3.19) disappears on integrating by parts, as we find from (3.15). We now move to the third term. Owing to (3.11) and (3.21), we may write

$$f_\mu = \frac{1}{2}\frac{\partial_\mu F}{F} = \frac{1}{2}\frac{F'}{F}\,\partial_\mu\phi, \tag{3.22}$$

where

$$F' = \frac{dF}{d\phi}.$$

In this way we have put (3.18) into the form

$$\mathcal{L}_1 = \sqrt{-g_*}\left(\frac{1}{2}R_* - \frac{3}{4}\left(\frac{F'}{F}\right)^2 g_*^{\mu\nu}\,\partial_\mu\phi\,\partial_\nu\phi\right). \tag{3.23}$$

As we stated before, the first term is of the form of the Einstein–Hilbert term but without nonminimal coupling. Also the second term looks similar to the kinetic term of ϕ.

By the same procedure, the second term of (3.16) can be put into the form

$$-\tfrac{1}{2}\sqrt{-g}\epsilon g^{\mu\nu}\,\partial_\mu\phi\,\partial_\nu\phi = -\tfrac{1}{2}\sqrt{-g_*}\epsilon\Omega^{-2}g_*^{\mu\nu}\,\partial_\mu\phi\,\partial_\nu\phi$$

$$= -\tfrac{1}{2}\sqrt{-g_*}\epsilon\frac{1}{F}g_*^{\mu\nu}\,\partial_\mu\phi\,\partial_\nu\phi, \tag{3.24}$$

where (3.20) has been used. We find that (3.24) has the same appearance as the second term of (3.23). Collecting them together yields

$$-\tfrac{1}{2}\sqrt{-g_*}\Delta g_*^{\mu\nu}\,\partial_\mu\phi\,\partial_\nu\phi, \tag{3.25}$$

where

$$\Delta = \frac{3}{2}\left(\frac{F'}{F}\right)^2 + \epsilon\frac{1}{F}. \tag{3.26}$$

If

$$\Delta > 0, \tag{3.27}$$

we introduce a function $\sigma(x)$ that satisfies

$$\frac{d\sigma}{d\phi} = \sqrt{\Delta}. \tag{3.28}$$

We then find

$$\sqrt{\Delta}\,\partial_\mu\phi = \frac{d\sigma}{d\phi}\,\partial_\mu\phi = \partial_\mu\sigma,$$

which allows us to re-express (3.25) as

$$-\tfrac{1}{2}\sqrt{-g_*}\,g_*^{\mu\nu}\,\partial_\mu\sigma\,\partial_\nu\sigma, \tag{3.29}$$

a *canonical* kinetic term of the new scalar field σ. In other words, σ instead of the original ϕ behaves as the usual scalar field in the new conformal frame.

If $\Delta < 0$, the sign in front of (3.29) is plus, implying that σ is a ghost. In order for σ to be a normal field with a positive energy, the condition (3.27) has to be satisfied. This can be achieved even if $\epsilon < 0$, as was remarked before.

The relation between σ and ϕ can be obtained by integrating (3.28). For an explicit form (3.17), we find

$$\Delta = \left(6 + \epsilon\xi^{-1}\right)\phi^{-2} = \zeta^{-2}\phi^{-2}, \tag{3.30}$$

where ζ is defined by (2.21). The condition $\Delta > 0$ is then given by

$$\zeta^2 > 0, \tag{3.31}$$

which agrees with (2.56), suggesting that σ introduced in Chapter 2 is the same as the one considered in this chapter. We emphasize, however, that the previous discussion was limited to the weak-field approximation, whereas the present argument is true in a fully general situation.

We now substitute (3.30) into (3.28), obtaining

$$\frac{d\sigma}{d\phi} = \zeta^{-1}\phi^{-1}, \tag{3.32}$$

which is integrated to yield

$$\zeta\sigma = \ln\left(\frac{\phi}{\phi_0}\right), \qquad \text{or} \qquad \phi = \phi_0 e^{\zeta\sigma}, \tag{3.33}$$

where an integration constant ϕ_0 is to be determined later.

In terms of this σ, (3.16) has finally been put into the form

$$\mathcal{L}_{\mathrm{BD}} = \sqrt{-g_*}\left(\tfrac{1}{2}R_* - \tfrac{1}{2}g_*^{\mu\nu}\,\partial_\mu\sigma\,\partial_\nu\sigma + L_{*\mathrm{matter}}\right). \tag{3.34}$$

This is the Lagrangian in the new conformal frame, where $L_{*\mathrm{matter}}$ is the conformally transformed matter Lagrangian, which will be discussed in detail in the next section. Before doing that we make some remarks.

It seems convenient to give names to each of the conformal frames before and after the conformal transformation determined by (3.20). We propose to call them the J frame and the E frame, respectively. "J" comes from Jordan, who should be remembered to have discussed nonminimal coupling for the first time. By "E" we imply a conformal frame in which there is no nonminimal coupling as in Einstein's theory. Equation (3.14) is the Lagrangian expressed in the J frame, whereas the "same" theory is defined by the Lagrangian (3.34) if it is viewed in the E frame.

We have chosen Ω by applying the condition (3.21). Obviously, however, we are free to choose other $\Omega(x)$. In this sense there is an infinite number of conformal frames, among which the J and E frames have "privileged" status. We learned that the presence or absence of the nonminimal coupling term, and its form, depend on the choice of a conformal frame. It may appear that a theory is "trivial" in the E frame because we have Einstein's theory with an additional scalar field σ. We point out, however, that the scalar field now occurs in the matter Lagrangian $L_{*\mathrm{matter}}$ in a specific manner that reflects what the theory looks like in other conformal frames.

It may also appear that the gravitational constant is a true constant, and hence that there is no time-variability in cosmology, because there is no nonminimal coupling in the E frame. This seems to imply that physical conclusions depend on the choice of conformal frames, raising the question of whether there is any unambiguous principle by which we can select a conformal frame uniquely. We are going to discuss the issue as we extend the argument to explicit examples.

We should also add that $\Delta > 0$, the condition that σ is not a ghost, can be satisfied even with $\epsilon = -1$, implying that ϕ is "apparently" a ghost. We only require

$$\xi > \tfrac{1}{6}. \tag{3.35}$$

The underlying reason can be understood if we look at the content of (3.25) or (3.26). In fact the first term of (3.26) comes from the nonminimal coupling term, which is always positive, the sign appropriate for σ to be a nonghost, a normal field. On the other hand, the second term is a contribution from the usual kinetic term, which may have the "wrong" sign but can be compensated if the first term is large enough. As we learned from the discussion in the preceding chapter, this is the effect of mixing with the spinless component of the metric field, though it is not *a priori* obvious why we finally come to a "right" sign; this is a nontrivial aspect of scalar–tensor theories.

One might even choose $\epsilon = 0$, still reaching $\zeta^2 = \tfrac{1}{6} > 0$, and hence giving a positive energy for σ. In fact an interesting model of this type according to which the scalar field acquires a dynamical degree of freedom even without a kinetic term in the J frame, as a consequence purely of mixing, was proposed.

One might argue that, with $\epsilon = -1$, the ϕ field should yield a negative energy in flat space-time in the J frame, which is an outright difficulty [22]. It is crucially important in this connection, however, to recall that ϕ and the spin-0 component of the metric field are mixed with each other as a consequence of the mixing interaction. It then follows that any excitation in the ϕ mode should entail excitation in the metric mode, not allowing space-time to stay flat. According to the previous discussion, the total energy of the whole system should be kept positive. We admit that ϕ as part of matter would also affect space-time geometry, but this is certainly a higher-order effect. In the E frame, on the other hand, the situation is simpler; at least to the lowest nontrivial order, σ is no longer mixed; hence it is allowed to be excited with positive energy, leaving space-time flat.

The integration constant ϕ_0 introduced in (3.33) can be determined in the following manner. Start by considering that $\phi = \phi_0$ corresponds to $\sigma = 0$, for which F is constant, and hence there is no need for a conformal transformation. We may then set $\Omega = 1$. Combine this with the fact that we fixed Ω by the condition (3.20), obtaining

$$\xi\phi_0^2 = 1, \tag{3.36}$$

from which it follows that

$$\phi_0 = \xi^{-1/2}. \tag{3.37}$$

In this way we may put (3.33) into the form

$$\phi = \xi^{-1/2}e^{\zeta\sigma}. \tag{3.38}$$

We expand this into a series

$$\phi = \xi^{-1/2} + \xi^{-1/2}\zeta\sigma + \cdots, \tag{3.39}$$

which agrees with (2.24) for which we use (2.24) and (2.118).

We then have (3.21) in the form

$$\Omega = e^{\zeta\sigma}, \tag{3.40}$$

which is substituted into (3.1). The result can be expressed in a linearized form;

$$g_{*\mu\nu} = \eta_{\mu\nu} + h_{*\mu\nu}, \tag{3.41}$$

where

$$h_{*\mu\nu} = h_{\mu\nu} + 2\eta_{\mu\nu}\zeta\sigma. \tag{3.42}$$

As a standard procedure we introduce

$$\chi_{*\mu\nu} \equiv h_{*\mu\nu} - \tfrac{1}{2}\eta_{\mu\nu}h_*, \tag{3.43}$$

which is computed to be

$$\chi_{*\mu\nu} = h_{\mu\nu} - \tfrac{1}{2}\eta_{\mu\nu}h - 2\eta_{\mu\nu}\zeta\sigma, \tag{3.44}$$

in agreement with (2.42) with the condition (2.118).

From these considerations we find that the approach in terms of a conformal transformation developed in this chapter has an aspect of generalizing diagonalization in the linearized theory discussed in Chapter 2. It is remarkable to find that the same coefficient ζ which appears in (3.30) in the general theory had occurred also in the linearized version.

In some physical applications, as we know, the linear approximation is sufficient, and hence the method of conformal transformations might not be particularly useful. As an exception, quantum corrections with respect to coupling to matter can usually be calculated in terms of the interaction Hamiltonian which exists only in the E frame. We also add a warning that naive and intuitive arguments might not be justified in conformal frames in which nonminimal coupling is present, as will be demonstrated in Chapter 5.

We close this section by studying another extreme choice with $\epsilon = -1$, $\xi = \tfrac{1}{6}$. From (3.30) we obtain $\Delta = 0$, which implies, as we find from (3.25), that there is no kinetic term for the scalar field in the E frame;

$$\mathcal{L}_{\mathrm{BD}} = \sqrt{-g_*}\left(\tfrac{1}{2}R_* + L_{*\mathrm{matter}}\right). \tag{3.45}$$

There is no dynamical degree of freedom of the scalar field. This particular example allows us, however, to consider a somewhat different interpretation, as we discuss briefly in Appendix H.

3.3 Coupling to matter

In section 3.2 we derived (2.20) as a field equation in the J frame representing coupling to matter of the scalar field:

$$\Box\varphi = \zeta^2 T, \tag{3.46}$$

which we will re-express in the E frame.

First the left-hand side can be put into the form

$$\Box\varphi = \frac{1}{2}\xi\frac{1}{\sqrt{-g}}\,\partial_\mu\left(\sqrt{-g}\,g^{\mu\nu}\,\partial_\nu\phi^2\right). \tag{3.47}$$

We use

$$\phi^2 = \xi^{-1} e^{2\zeta\sigma},$$
$$\sqrt{-g} = \sqrt{-g_*} e^{-4\zeta\sigma},$$
$$g^{\mu\nu} = g_*^{\mu\nu} e^{2\zeta\sigma}, \tag{3.48}$$

finding

$$\Box\varphi = \zeta e^{4\zeta\sigma} \Box_*\sigma. \tag{3.49}$$

On the right-hand side of (3.46), we recall (2.9);

$$T_{\mu\nu} = -2\frac{1}{\sqrt{-g}} \frac{\delta\mathcal{L}_{\text{matter}}}{\delta g^{\mu\nu}}. \tag{3.50}$$

We now apply a chain rule,

$$\frac{\delta\mathcal{L}_{\text{matter}}}{\delta g^{\mu\nu}} = \frac{\partial g_*^{\rho\sigma}}{\partial g^{\mu\nu}} \frac{\delta\mathcal{L}_{\text{matter}}}{\delta g_*^{\rho\sigma}} = \Omega^{-2} \frac{\delta\mathcal{L}_{\text{matter}}}{\delta g_*^{\mu\nu}}, \tag{3.51}$$

though, rigorously speaking, this can be verified only if $\mathcal{L}_{\text{matter}} \equiv \sqrt{-g}L_{\text{matter}}$ contains no derivative of $g^{\mu\nu}$, and hence a functional derivative with respect to $g^{\mu\nu}$ is reduced to a simple derivative. We notice, however, that we are finally interested only in T, which is part of the whole $T_{\mu\nu}$, hence staying practically under the above restriction. In this sense we have

$$T_{\mu\nu} = \Omega^2 T_{*\mu\nu}, \tag{3.52}$$

where $T_{*\mu\nu}$ is defined by (3.50) with $*$ attached everywhere. We recall, however, that $\mathcal{L}_{*\text{matter}}$ is the same as $\mathcal{L}_{\text{matter}}$ but re-expressed in terms of quantities with $*$.

In this way we have

$$T = g^{\mu\nu}T_{\mu\nu} = \Omega^4 T_*. \tag{3.53}$$

Substituting this together with (3.48) into (3.46) now yields

$$\Box_*\sigma = \zeta T_*, \tag{3.54}$$

implying that σ has its source given also by the trace of the matter energy–momentum tensor. Combining this with the fact that σ is no longer mixed with the metric tensor in the E frame, we must conclude that σ has direct coupling to matter in the Lagrangian. We will see how this is realized in the following.

For this purpose, we consider a real scalar field Φ as a simplified representative of matter fields. This scalar field should not be confused with the gravitational scalar field, ϕ, σ, or χ.

Let us assume the matter Lagrangian of a free real massive scalar field,

$$\mathcal{L}_{\text{matter}} = \sqrt{-g}\left(-\tfrac{1}{2}g^{\mu\nu}\,\partial_\mu\Phi\,\partial_\nu\Phi - \tfrac{1}{2}m^2\Phi^2\right), \tag{3.55}$$

where no coupling to ϕ is introduced.

We apply the conformal transformation (3.6):

$$g_{\mu\nu} = \Omega^{-2}g_{*\mu\nu}, \tag{3.56}$$

with (3.40):

$$\Omega = \xi^{1/2}\phi = e^{\zeta\sigma}, \tag{3.57}$$

to re-express the Lagrangian (3.55) as

$$\mathcal{L}_{\text{matter}} = \sqrt{-g_*}\left(-\tfrac{1}{2}\Omega^{-2}g_*^{\mu\nu}\,\partial_\mu\Phi\,\partial_\nu\Phi - \tfrac{1}{2}\Omega^{-4}m^2\Phi^2\right). \tag{3.58}$$

The kinetic term can be made canonical in terms of a new field Φ_* defined by

$$\Phi = \Omega\Phi_*, \tag{3.59}$$

thus further re-expressing (3.58) as

$$\mathcal{L}_{\text{matter}} = \sqrt{-g_*}\left(-\tfrac{1}{2}g_*^{\mu\nu}\,\mathcal{D}_\mu\Phi_*\,\mathcal{D}_\nu\Phi_* - \tfrac{1}{2}m_*^2\Phi_*^2\right), \tag{3.60}$$

where

$$\mathcal{D}_\mu\Phi_* = [\partial_\mu + \zeta\,(\partial_\mu\sigma)]\Phi_*, \tag{3.61}$$
$$m_*^2 = \Omega^{-2}m^2, \tag{3.62}$$

or

$$m_* = \Omega^{-1}m = me^{-\zeta\sigma}. \tag{3.63}$$

The matter energy–momentum tensor can be derived from (3.60):

$$T_{*\mu\nu} = -2\frac{1}{\sqrt{-g_*}}\frac{\partial\mathcal{L}_{\text{matter}}}{\partial g_*^{\mu\nu}}$$
$$= \mathcal{D}_\mu\Phi_*\,\mathcal{D}_\nu\Phi_* + g_{*\mu\nu}\left(-\tfrac{1}{2}g_*^{\mu\nu}\,\mathcal{D}_\mu\Phi_*\,\mathcal{D}_\nu\Phi_* - \tfrac{1}{2}m_*^2\Phi_*^2\right). \tag{3.64}$$

The field equation of Φ_* is derived as

$$\left(\Box_* - m_*^2\right)\Phi_* + \zeta\Phi_*\,\Box_*\Phi_* - \zeta^2(\partial\sigma)^2\Phi_* = 0. \tag{3.65}$$

By substituting (3.60) into (3.34), we also obtain the equation for σ:

$$\left(1 + \zeta^2\Phi_*^2\right)\Box_*\sigma + \zeta m_*^2\Phi_*^2 + \tfrac{1}{2}\zeta\,\Box_*\left(\Phi_*^2\right) + \zeta^2(\partial\sigma)\left(\partial\Phi_*^2\right) = 0. \tag{3.66}$$

On combining these equations we finally obtain

$$\Box_* \sigma = \zeta T_*, \tag{3.67}$$

where T_* is the trace of $T_{*\mu\nu}$ obtained in (3.64). In this way we have recovered (3.54) as an explicit consequence of the fact that σ appears in the matter Lagrangian in the E frame.

Complications have occurred in (3.64)–(3.66) due to the presence of the term in $\partial_\mu \sigma$ in (3.61). No extra term of this kind occurs if we consider a spinor field, resulting much more easily in the same result (3.67), as will be shown in Appendix F. This indicates that (3.67) is a general result independent of the content of the matter energy–momentum tensor.

Notice that (3.54) looks the same as (2.119) derived in the linear approximation in the sense that the coefficient on the right-hand side is ζ.

Before closing this section, however, it seems appropriate to add a remark. We have mainly discussed conformal transformation from the J frame to the E frame. We may also consider a transformation in the reverse direction, by introducing ϕ by (3.33) and reproducing the "original" Lagrangian (3.16) in terms of ξ defined by

$$\xi^{-1} = \epsilon\left(\zeta^{-2} - 6\right). \tag{3.68}$$

One may notice here that no ζ is found in (3.34) except in $L_{*\text{matter}}$. In this context ξ can be anything. It seems to follow that, starting from a J-frame Lagrangian with a certain value of ξ, and moving to an E frame, one may go to *another* J frame with a different value of ξ, although ϕ is not the same as before. Can one conclude that the value of ξ has no real meaning?

The answer is no, because $L_{*\text{matter}}$ depends on ζ. This might already have been suggested in (3.54), but we will show an explicit example in (F.26) in Appendix F. We remind readers that ζ or ξ measures how strongly the scalar field is coupled to matter. It does not make sense if we ignore the matter sector, focusing only on the gravity sector including the scalar field.

3.4 The geodesic in the E frame

We recall, however, that the absence of a direct coupling to matter of the scalar field in the J frame assures the validity of the WEP, and may ask what the corresponding situation in the E frame is. This is also related to the fact that, owing to this coupling in the E frame, the covariant conservation law is now broken, and hence the right-hand side of the equation of motion of a particle, the geodesic equation multiplied by mass, no longer vanishes. The right-hand side is, however, precisely proportional to mass, as we find from (F.37) in Appendix F.

The coefficient is ζ, common to any particle. By dividing both sides by mass, we obtain an equation totally independent of mass, thus maintaining the WEP unspoiled.

The same conclusion can be confirmed also by transforming a geodesic equation directly. To see this we use (3.10) to re-write the geodesic equation in the J frame,

$$\frac{Du^{\mu}}{D\tau} \equiv \frac{du^{\mu}}{d\tau} + \Gamma^{\mu}{}_{\nu\lambda}u^{\nu}u^{\lambda} = 0. \tag{3.69}$$

From (3.2) we have

$$\frac{d\tau_{*}}{d\tau} = \Omega, \tag{3.70}$$

from which it follows that

$$u^{\mu} = \frac{dx^{\mu}}{d\tau} = \Omega u_{*}^{\mu}, \tag{3.71}$$

and hence

$$\frac{du^{\mu}}{d\tau} = \frac{d\tau_{*}}{d\tau}\frac{d}{d\tau_{*}}(\Omega u_{*}^{\mu}) = \Omega^{2}\left(\frac{du_{*}^{\mu}}{d\tau_{*}} + f_{\nu}u_{*}^{\nu}u_{*}^{\mu}\right), \tag{3.72}$$

where we used

$$\frac{d\Omega}{d\tau_{*}} = \frac{\partial\Omega}{\partial x^{\nu}}\frac{dx^{\nu}}{d\tau_{*}}.$$

Substituting (3.10) into this yields

$$\frac{Du^{\mu}}{D\tau} = \Omega^{2}\left(\frac{Du_{*}^{\mu}}{D\tau_{*}} - (f_{\nu}u_{*}^{\nu})u_{*}^{\mu} + f_{*}^{\mu}(u_{*\nu}u_{*}^{\nu})\right). \tag{3.73}$$

We then put (3.69) into the form

$$\frac{Du_{*}^{\mu}}{D\tau_{*}} = (f_{\nu}u_{*}^{\nu})u_{*}^{\mu} - f_{*}^{\mu}(u_{*\nu}u_{*}^{\nu}). \tag{3.74}$$

Using (3.40), we finally obtain

$$\frac{Du_{*}^{\mu}}{D\tau_{*}} = \zeta\left[\left(u_{*}^{\lambda}\,\partial_{\lambda}\sigma\right)u_{*}^{\mu} - g_{*}^{\mu\lambda}\,\partial_{\lambda}\sigma(u_{*\nu}u_{*}^{\nu})\right]. \tag{3.75}$$

The right-hand side is nonvanishing, but contains σ, a field common to all the particles, independent of specific properties of particles, such as mass. A geodesic equation is a sufficient but not a necessary condition for the validity of the WEP.

After all, if different objects are assured to move in the same way in a conformal frame, the same should be true in any other conformal frame.

In Appendix F, we derive (F.27):

$$\left(\not{D}_* + me^{-\zeta\sigma} \right)\psi_* = 0, \tag{3.76}$$

as a Dirac equation in the E frame. On the other hand, the energy–momentum tensor of this field is determined by the kinetic term, which shares the same form in the J frame (with $*$ attached everywhere). The trace is calculated by using an equation of motion giving

$$T_* = \overline{\psi}_* \not{D}_* \psi_* = -me^{-\zeta\sigma}\overline{\psi}_* \psi_*. \tag{3.77}$$

Equation (F.26) in Appendix F gives a source of σ, yielding the field equation of σ,

$$\Box_* \sigma = -\zeta me^{-\zeta\sigma}\overline{\psi}_* \psi_* = \zeta T_*, \tag{3.78}$$

with the same form as the corresponding (2.119), which is also in agreement with (3.54).

4

Cosmology with Λ

After section 4.1 giving a brief history of the problem of the cosmological constant, we go up the ladder starting from the standard theory with Λ added (section 4.2), proceeding to the prototype BD model without Λ (section 4.3), and culminating in the prototype BD model with Λ included (section 4.4), where the discussion will concern both the J frame and the E frame. We will face some crucial aspects that Λ has brought into being for the first time. Most remarkable is that the attractor solution in the J frame represents a static universe. This conclusion turns out to be evaded in the E frame, but particle masses are shown to vary with time too much. We then propose in subsection 4.4.3 a remedy in the matter part of the Lagrangian, thus violating the WEP in a manner that, we hope, allows us to remain within the observational constraint. At this cost, however, we are rewarded with a successful implementation of the scenario of a decaying cosmological constant in the E frame, which is now considered to be (approximately) physical. Another point to be noticed is that a physical condition, positivity of the energy density of matter, requires that $\epsilon = -1$, an apparently ghost nature of the scalar field in the J frame, unexpectedly in accordance with what string theory and KK theory suggest. This also entails the condition $\xi > \frac{1}{6}$, and thus is in contradiction with the widely known constraint $\omega \gtrsim 3.6 \times 10^3$, or $\xi \lesssim 7.0 \times 10^{-5}$. A reconciliation with the solar-system experiments will be made only with a nonzero mass of the scalar field.

4.1 How has the cosmological constant re-emerged?

One of the aims that the pioneers of the scalar–tensor theory expected to achieve was to prepare a theory that makes it possible to implement the idea of a gravitational constant that varies with cosmic time. In this sense, cosmology was among the high-priority list of applications at the outset. Since VLBI finally established the bound $\omega \gtrsim 3.6 \times 10^3$, or $\xi \lesssim 7.0 \times 10^{-5}$, the prototype BD model seems to have been destined to be ever more a minority interest in the physics of gravity. Only the proposal of non-Newtonian gravity featuring a macroscopic force-range appeared to suggest a new direction for the scalar–tensor theory free from the above-mentioned constraint on the parameter, though still without any solid evidence.

For the last few decades or so, on the other hand, the cosmological constant has appeared increasingly to be a real issue as the measurement technology has been improved, enhancing the expectation for something beyond the standard theory. Before going into the details of this most up-to-date topic, however, it seems appropriate here to begin with a short history of the problem of the cosmological constant.

Einstein noticed that his gravitational field equation allowed the presence of a constant, denoted by Λ, which could not be ruled out theoretically, but was left unrelated to any hitherto-known observation. From his sense of simplicity and beauty, he decided to ignore this term. In 1922, however, A. Friedmann showed that the cosmological solution of Einstein's equation without Λ admitted only the possibilities of the universe expanding or contracting, a conclusion totally unacceptable even to Einstein, who had been so radical in bringing new ideas into physics. He might have been uneasy unless the whole universe, the very foundation of everything, was assured to remain stubbornly unchanging. In order to have the solution of a static universe, he introduced Λ very reluctantly.

On reading the report a few years later on E. Hubble's discovery that the universe was expanding, Einstein declared immediately that he had made the biggest blunder in his life by calling for Λ. It appeared as if Λ had been expelled from physics once and for all. More than half a century later, however, we find ourselves haunted by the ghost of Λ still roaming around like an uninvited guest at our doors, this time with two faces; why is it so small and why does it exist at all? These issues will be discussed in the following.

From a theoretical point of view, unified theories make it ever more unavoidable to have Λ of the order of $M_{\rm P}^4$, as large as 120 orders of magnitude greater than the observational upper bounds. There was a period of several years during which there was much effort devoted to trying to explain how one can remove Λ altogether without appealing to the unnatural, unrealistic, and unimaginable accuracy of 120 orders of magnitude, called a "fine-tuning problem," arising basically from the "hierarchy problem." Most symmetry arguments failed, as was portrayed in [16]. It also seems unlikely that the topological argument at the level of quantum cosmology [60] is directly relevant to the late universe.

One might argue that supersymmetry forbids quantum-theoretical vacuum energy, which acts as a cosmological constant, being nonvanishing. This does not help in practice in a number of ways. First, it was suggested that, from a supersymmetric point of view, a cosmological term is not a pure constant, but is a field proportional to $\sqrt{-g}$, which can be a member of a super-multiplet [61]. We also know of several different ways in which a four-dimensional Λ may emerge, even starting with more fundamental theories without a cosmological constant in higher dimensions. KK theory in six dimensions, for example, results in Λ when two-dimensional space is compactified, as is explained in Appendix A. The conclusion remains essentially the same for any other dimensionality, as discussed in Appendix K.

Also in the 11-dimensional theory, the third-rank antisymmetric tensor field may acquire a nonzero vacuum expectation value of the field strength, $F_{\mu\nu\rho\sigma} = \text{constant} \times \epsilon_{\mu\nu\rho\sigma}$, thus providing another example of a four-dimensional Λ, represented by $\sqrt{-g}F_{\mu\nu\rho\sigma}F^{\mu\nu\rho\sigma} = \sqrt{-g} \times \text{constant}$.

Also important is to notice that supersymmetry is broken at the mass scale of the order of $M_{\text{ssb}} \sim 1\,\text{TeV}$, thus leaving a nonzero vacuum energy density of the order of $\sim(1\,\text{TeV})^4$. As another source of Λ, this is still too large by a factor of $\sim 10^{60}$. All in all there seems to be no way in which one may comfortably forget about a cosmological constant if we believe in the idea of unification.

In the above arguments, we recognize that there are two origins that might result in a cosmological constant. One is what we expect from M_{P}^4, which probably came into being near the Planck time, and left an effect as a classical constant, whereas the other, given above by $\sim M_{\text{ssb}}^4$, comes from a quantum fluctuation that is taking place even at this moment. The difference between the two might be compared to the relation between a prehistoric legacy and daily occurrences. The scenario of a decaying cosmological constant, which will be discussed in later sections, will apply only to Λ coming from the classical origin. The one with the quantum origin seems more difficult to deal with, though we offer a conjecture near the end of subsection 4.4.2.

We also add a comment that the theoretical idea of a scalar field in cosmology has its own history, which is not necessarily connected with the scalar–tensor theory. In less constrained frameworks, there has been an expectation that a massless or nearly massless scalar field might provide us with a solution to the problem of the cosmological constant, at least in a classical sense [62].

In addition to recognizing the two origins as mentioned above, we are going to point out that today's version of the problem of the cosmological constant shows two faces, as well. The issues discussed above are rather traditional, concerned mainly with how Λ can be made to vanish or be infinitesimally small. This might be viewed as the first face of the problem of the cosmological constant, which was discussed during the first phase, which may be dated back to the advent of the modern versions of unified theories, before the latest decade when we began to glimpse the possibility of there being a small but nonzero cosmological constant within the upper bound established before. It seems better to recognize the latter feature as the second face the problem shows. Details on the recent developments during what we call the second phase of the contemporary version of the problem of the cosmological constant will be the main topic in Chapter 5.

A cosmological constant is looming in yet another context of contemporary cosmology. It is widely accepted that, in the very early stage of development of the universe, an extremely rapid expansion is supposed to have taken place, presumably exponentially, like the one caused by Λ. Originally this "inflation" era was considered to be driven by the vacuum energy [63–65], whereas the later versions, a "new inflation" [66] or a "chaotic inflation" [67], made use of a classical scalar field falling down a potential slope. A large term characterized essentially by M_P will give the required effect, not necessarily calling for a "constant."

However, there still remains the question of how this inflationary behavior came to an end. The eternal expansion implied by a true constant is totally unrealistic. A nonconstant potential of a scalar field may appear to be free from this difficulty *only if* the field settled at a minimum which is zero. If the minimum is nonzero, on the other hand, it would have played the same role as a nonzero cosmological constant. One must adjust the bottom of the potential to zero unrealistically accurately. In this sense the successful idea of inflation rests again on the solution of the problem of the cosmological constant.

Some of the details of the inflation mechanism, however, still remain unsolved. In particular, the matter density must have been "re-heated" after it had redshifted considerably during the period of rapid expansion, to the extent that it recovered a temperature sufficiently high to lead to the development that follows, i.e. the GUT epoch. This requires the satisfaction of some conditions that seem to go beyond the capabilities of the simple models. For this reason, we limit our considerations mainly to the time after the phase of re-heating; we start at the initial time $\sim 10^{10}$ in units of the Planck time, for example, to integrate the cosmological equations.

In section 4.2, we recall how a cosmological constant, if there is one, would affect the conventional theory. We sketch the primordial inflationary scenario, which has become an indispensable ingredient of contemporary cosmology. We then move on in section 4.3 to revisit briefly the prototype BD model *without* Λ, followed by section 4.4 in which we analyze the same model but now *with* Λ. We apply the model to the spatially flat Robertson–Walker universe. We find that the model does implement the scenario of "a decaying cosmological constant," but, rather unexpectedly, entails a background cosmological evolution vastly different from what one expects naturally in the standard cosmology. This is shown both in the J frame and in the E frame. A drastic modification appears to be called for. One may start by modifying the model by changing the form of the nonminimal coupling term gradually, for example. In view of the fact that the disagreement is quite dramatic, however, we prefer a departure in a decisive manner. Rather surprisingly, we find that the required change in the result would come most easily from one in the coupling to matter instead of one in the gravitational portion itself.

In this context subsection 4.3.3 will be devoted to our proposal to introduce a coupling to matter of the scalar field, going against one of the premises in the prototype BD model, and thus violating the WEP. We show that a coupling that respects dilatation invariance in four dimensions is favored, also suggesting a mechanism that provides a common origin of the gravitational constant and particle masses [68]. Apart from this esthetic feature, we find the rather unexpected result that the scalar field is now decoupled from matter in the E frame. For this reason particle masses stay completely time-independent, hence leaving the E frame fully physical; meter-sticks and clocks made up basically of elementary particles are truly constant. The WEP, which is now violated in principle, shows up hardly at all as the effect of forces between objects. It also turns out, however, that this "desired" feature no longer persists if quantum effects among matter fields are taken into account, as will be discussed in Chapter 6. Until then we stay in the classical arena, because the effects are not overwhelming.

We add that, even with this modification, the physical condition of positivity of the matter density requires the condition $\epsilon = -1$, an apparent ghost nature of ϕ in the J frame, in accordance with what KK theory as well as the string dilaton suggest. An overall positive energy can be achieved only if $\xi > \frac{1}{6}$, which does not allow the scalar–tensor theory to go to the limit of the standard theory, thus enhancing the status of the theory beyond that of a model meant to describe an infinitesimal deviation from the standard theory. This also forces us to reconsider the constraint $\xi \lesssim 7.0 \times 10^{-5}$. A reconciliation with the solar-system experiments will be achieved by assuming a nonzero finite mass of the scalar field, a subject in Chapter 6 where quantum effects are taken into account.

4.2 The standard theory with Λ

It appears as if we have no other way but to live peacefully with Λ. We should be prepared for what "danger" it might have potentially. For this purpose we start by considering the standard theory but simply by Λ added;

$$\mathcal{L}_\Lambda = \sqrt{-g}\left(\tfrac{1}{2}R - \Lambda + L_{\mathrm{m}}\right). \tag{4.1}$$

The effect coming from Λ may be viewed as an additional contribution to the energy–momentum tensor represented by

$$T^{(\Lambda)}_{\mu\nu} = -2\frac{1}{\sqrt{-g}}\frac{\partial}{\partial g^{\mu\nu}}\left(-\sqrt{-g}\Lambda\right) = -g_{\mu\nu}\Lambda. \tag{4.2}$$

This part is covariantly conserved by itself.

As a first approximation in cosmology we as usual choose a spatially uniform, isotropic Robertson–Walker metric. We further assume that we have a spatially flat universe; hence

$$g_{00} = -1, \qquad g_{ij} = a^2(t)\delta_{ij}, \qquad g_{0i} = 0, \tag{4.3}$$

where $x^0 = t$ is the cosmic time, while $a(t)$ is the scale factor of the universe. The ordinary matter energy–momentum tensor has then the components

$$T_{00} = \rho, \quad T_{ij} = Pg_{ij} = Pa^2\delta_{ij}. \tag{4.4}$$

On comparing (4.2) and (4.4) we find that Λ contributes additional energy density and pressure;

$$\rho_\Lambda = \Lambda \quad \text{and} \quad P_\Lambda = -\Lambda. \tag{4.5}$$

Note that a positive Λ acts as if it provided a fluid of *negative* pressure.

The 00-component of Einstein's equation is now

$$3H^2 = \rho + \Lambda, \tag{4.6}$$

where $H = \dot{a}/a$.

From the covariant conservation law of $T_{\mu\nu}$ we derive as usual

$$\rho = 3Ca^{-3(1+z)}, \tag{4.7}$$

where we assumed the equation of state, $P = z\rho$, while C is a positive constant. Using this in (4.6) and expecting that the universe is expanding instead of contracting, we obtain

$$\frac{\dot{a}}{a} = \sqrt{Ca^{-3(1+z)} + \frac{\Lambda}{3}}. \tag{4.8}$$

Consider first the situation in which

$$\rho \gg \Lambda. \tag{4.9}$$

Then solving (4.8) yields a standard power-law expansion;

$$a(t) = a_0 t^\gamma, \tag{4.10}$$

with

$$\gamma = \frac{2}{3}\frac{1}{1+z}. \tag{4.11}$$

On substituting (4.10) into (4.7) we also find

$$\rho(t) \sim t^{-2}. \tag{4.12}$$

Owing to this decrease, the universe will eventually enter the Λ-dominated epoch;

$$\rho(t) \ll \Lambda. \tag{4.13}$$

Ignoring the first term inside the square root in (4.8) gives

$$H = \sqrt{\frac{\Lambda}{3}}, \tag{4.14}$$

from which it follows that

$$a(t) = a_1 \exp\left(\sqrt{\frac{\Lambda}{3}}t\right), \tag{4.15}$$

representing an exponential growth of the scale factor, or "inflation."

No such rapid expansion is indicated at present, or throughout almost the whole history of the universe, except during the conjectured primordial inflationary period. As of today, particularly, we are not in the Λ-dominated epoch. This implies that, on the right-hand side of (4.6),

$$\Lambda \lesssim \rho \sim \rho_{\text{cr}}, \tag{4.16}$$

where the "critical density" today is given by

$$\rho_{\text{cr}} = 3H_0^2 \approx \frac{4}{3}t_0^{-2}. \tag{4.17}$$

Here the present age $t_0 \approx 10^{10}$ years is about 10^{60} in units of Planck time. In this way we arrive easily at an upper bound

$$\Lambda \lesssim 10^{-120}. \tag{4.18}$$

In other words, today's Λ is smaller than the theoretically natural value ~ 1 by as much as 120 orders of magnitude, a widely appreciated number.

4.3 The prototype BD model without Λ

Before going into the subject of cosmology based on the prototype BD model with Λ, it might be helpful if we briefly outline the basic cosmological results of the prototype BD model without Λ.

For this purpose we first study the *dust*-dominated universe for the spatially flat Robertson–Walker metric in the J frame. The equations are

$$6\varphi H^2 = \tfrac{1}{2}\epsilon\dot{\phi}^2 + \rho - 6H\dot{\varphi}, \qquad (4.19)$$

$$\ddot{\varphi} + 3H\dot{\varphi} = \zeta^2\rho, \qquad (4.20)$$

$$\dot{\rho} + 3H\rho = 0. \qquad (4.21)$$

Assume first the power-law behavior for the density;

$$\rho = \rho_0 t^\beta. \qquad (4.22)$$

Substituting this into (4.21) gives uniquely

$$H = -\frac{\beta}{3}t^{-1}. \qquad (4.23)$$

Inspecting (4.20) suggests

$$\dot{\varphi} = At^{\beta+1}, \qquad (4.24)$$

where A and β are constants. Integrating this yields

$$\varphi = \frac{A}{\beta+2}t^{\beta+2}. \qquad (4.25)$$

This diverges , however, for $\beta \to -2$, which will be found to be a standard result. To avoid this we write

$$\frac{A}{\beta+2} = \varphi_0, \qquad (4.26)$$

with a new constant φ_0, which is expected to be finite for $\beta \to -2$. In this sense we replace (4.25) by

$$\varphi = \varphi_0 t^{\beta+2}. \qquad (4.27)$$

We now substitute this together with (4.23) and (4.22) into (4.20), obtaining

$$(\beta+2)[(\beta+1) - \beta]\varphi_0 = \zeta^2\rho_0, \qquad (4.28)$$

and hence

$$\rho_0/\varphi_0 = u\zeta^{-2}, \qquad (4.29)$$

where we change the "variable" β into

$$u = \beta + 2. \tag{4.30}$$

We next calculate (4.19). We find

$$6\varphi H^2 = \tfrac{2}{3}\varphi_0 \beta^2 t^\beta, \tag{4.31}$$
$$6\dot{\varphi} H = -2\varphi_0 u\beta t^\beta. \tag{4.32}$$

Also, from (1.9),

$$\varphi = \tfrac{1}{2}\xi\phi^2, \tag{4.33}$$

we have

$$\phi = \sqrt{2\xi^{-1}\varphi_0}\; t^{\beta/2+1}, \tag{4.34}$$

from which it follows that

$$\dot{\phi} = \tfrac{1}{2}\sqrt{2\xi^{-1}\varphi_0}\; ut^{\beta/2}, \tag{4.35}$$

and hence

$$K = \tfrac{1}{2}\epsilon\dot{\phi}^2 = \tfrac{1}{4}\epsilon\xi^{-1}\varphi_0 u^2 t^\beta. \tag{4.36}$$

Substituting (4.22), (4.31), (4.32), and (4.36) into (4.19) yields

$$\frac{\rho_0}{\varphi_0} = \frac{2}{3}\beta^2 - 2u\beta - \frac{1}{4}\epsilon\xi^{-1}u^2. \tag{4.37}$$

Equating this with (4.29), and using (1.14),

$$\zeta^{-2} = 6 + \epsilon\xi^{-1}, \tag{4.38}$$

we obtain

$$\frac{3}{4}\zeta^{-2} = \frac{F(u)}{u+4}, \tag{4.39}$$

where

$$F(x) = \frac{(u+4)^2}{2u}. \tag{4.40}$$

In this way we finally obtain

$$\zeta^{-2} = \tfrac{2}{3}(1 + 4u^{-1}), \tag{4.41}$$

which determines u uniquely in terms of ζ. Note that the limit $\zeta \to 0$ from general relativity corresponds to $u = 0$, or $\beta = -2$. Eliminating u in (4.29) with the help of (4.41) is shown to reproduce the BD result (63) in [5], if we use $\phi_{\mathrm{BD0}} = 16\pi\varphi_0$ and $\zeta^{-2} = 6 + 4\omega$.

From (4.27) we also find

$$\frac{\dot{G}}{G} = -\frac{\dot{\varphi}}{\varphi} = -ut^{-1}, \tag{4.42}$$

which shows that $u = 0$ corresponds to the standard result of a truly constant G. Otherwise, time-dependent G should ensue, as expected originally by Jordan. The expected behavior, $G \sim t^{-1}$, would result only for $u = 1$, and hence for a special value $\zeta = \sqrt{3/10} \approx 0.55$ from (4.41), which is quite far away from the upper bound obtained from the solar-system experiments.

Suppose, on the other hand, that we assume the widely accepted value $\omega \gtrsim 3.6 \times 10^3$, as given by (1.15), or $\zeta^{-2} \gtrsim 1.4 \times 10^4$. By substituting this into (4.41), we find

$$u \lesssim 2 \times 10^{-4}, \tag{4.43}$$

which is entirely outside the range that allows the LNH to be useful, but giving

$$\left| \frac{\dot{G}}{G} \right| \lesssim 10^{-14} \, \text{years}^{-1}. \tag{4.44}$$

We now turn to the *radiation*-dominated universe. Equation (4.19) remains the same as for dust-dominance, while (4.20) and (4.21) are replaced by

$$\ddot{\varphi} + 3H\dot{\varphi} = 0, \tag{4.45}$$
$$\dot{\rho} + 4H\rho = 0, \tag{4.46}$$

respectively.

We assume (4.22),

$$\rho = \rho_0 t^\beta, \tag{4.47}$$

as before. Substituting this into (4.46) gives

$$H = -\frac{\beta}{4} t^{-1}, \tag{4.48}$$

with a corresponding difference from (4.23).

Using this further in (4.45), we find

$$\dot{\varphi} = A t^{(3/4)\beta}, \tag{4.49}$$

and its integration

$$\varphi = \frac{A}{\frac{3}{4}\beta + 1} t^{(3/4)\beta + 1} + B, \tag{4.50}$$

with another constant B. It seems *a posteriori* that β can at most be -2, and hence $\frac{3}{4}\beta + 1 \leq -\frac{1}{2}$, implying that the term B, if it is nonzero, always dominates the whole contribution. In the prototype BD model *with* Λ, as we will see shortly, however, the constant corresponding to B will vanish because, as will be discussed next, the scalar field in the E frame will be driven to infinity due to the potential coming from Λ. In the present model without Λ, on the other hand, the scalar field in the E frame, and hence in the J frame, receives no force and may come to a stop somewhere, hence a finite B that controls the asymptotic behavior. For this reason choosing B is crucially important in the radiation-dominated universe.

We are then unable to obtain an exact solution, in contrast to the case of a dust-dominated universe. We seek a first-order perturbative solution for $|A/B| \ll 1$.

Substituting (4.50) into (4.33) yields

$$\phi \approx \sqrt{2\xi^{-1}B}\left(1 + \frac{A}{2B}\frac{1}{\frac{3}{4}\beta + 1}t^{(3/4)\beta + 1}\right), \tag{4.51}$$

from which it follows that

$$\dot{\phi} \approx \sqrt{2\xi^{-1}B}\frac{A}{2B}t^{(3/4)\beta}, \tag{4.52}$$

showing a major difference from (4.35). As a result we obtain

$$K \approx \frac{1}{4}\epsilon\xi^{-1}\frac{A^2}{B}t^{(3/2)\beta}, \tag{4.53}$$

in place of (4.36).

Corresponding to (4.31) we find

$$6\varphi H^2 = \frac{3}{8}B\beta^2 t^{-2} + \frac{3}{8}A\frac{\beta^2}{\frac{3}{4}\beta + 1}t^{(3/4)\beta - 1}, \tag{4.54}$$

while (4.32) is replaced by

$$6\dot{\varphi}H = -\frac{3}{2}A\beta t^{(3/4)\beta - 1}. \tag{4.55}$$

We then find

$$6\varphi H^2 + 6\dot{\varphi}H - K = \frac{3}{8}B\beta^2 t^{-2} + \frac{3}{8}A\beta\left(\frac{\beta}{\frac{3}{4}\beta + 1} - 4\right)t^{3/4\beta - 1}$$

$$-\frac{1}{4}\epsilon\xi^{-1}\frac{A^2}{B}t^{(3/2)\beta}. \tag{4.56}$$

In practice, we consider the range of $\beta \sim -2$, for which $\frac{3}{4}\beta - 1 \sim -2.5$, and $\frac{3}{2}\beta \sim -3$. In this sense, the three terms of (4.56) arrange themselves according to the decreasing order $t^{-2}, t^{-2.5}, t^{-3}$. The balance in (4.19) with $\rho_0 t^\beta$ should be valid for the highest-order term t^{-2}, thus giving $\beta = -2$. The balances for the other two terms are missing at this moment, but they are expected to be achieved by including other terms that must have been ignored so far. An example is to add lower-order terms to the right-hand side of (4.49) at the outset.

For the highest-order term we have

$$B = \tfrac{2}{3}\rho_0, \qquad (4.57)$$

for $\beta = -2$, leaving A still undetermined. In this sense we arrive at

$$\varphi = \tfrac{2}{3}\rho_0 - 2At^{-1/2}. \qquad (4.58)$$

We also find

$$\frac{\dot\varphi}{\varphi} \approx \frac{3A}{2\rho_0}t^{-3/2}, \qquad (4.59)$$

which should give a varying G with $\dot{G}/G = -\dot\varphi/\varphi$, showing that the simple behavior $G \sim t^\gamma$ is no longer to be expected.

4.4 The prototype BD model with Λ

As a continuation of the preceding section, we believe it justified to consider the prototype BD model with Λ added, as the next simplest model.

4.4.1 Solution in the J frame

Let us start with the following Lagrangian in the J frame first [69, 70]:

$$\mathcal{L}_{\text{BD}\Lambda} = \sqrt{-g}\left(\tfrac{1}{2}\xi\phi^2 R - \tfrac{1}{2}\epsilon g^{\mu\nu}\,\partial_\mu\phi\,\partial_\nu\phi - \Lambda + L_{\text{matter}}\right). \qquad (4.60)$$

We expect $\Lambda \sim 1$ in accordance with many models of unified theories. If $\Lambda < 0$, the universe would have collapsed within a very short time. In order to have the universe that has survived for a long time as we see it, we confine ourselves to $\Lambda > 0$.

For the matter part we assume the conservation law

$$\nabla_\mu T^{\mu\nu} = 0, \qquad (4.61)$$

whatever its content might be. We re-iterate here that this is a consequence of the BD requirement that the scalar field is decoupled from matter in the J-frame Lagrangian.

The field equation (2.6) is supplemented by the term coming from Λ:

$$2\varphi G_{\mu\nu} = T_{\mu\nu} + T^{\phi}_{\mu\nu} - g_{\mu\nu}\Lambda - 2(g_{\mu\nu}\Box - \nabla_\mu\nabla_\nu)\varphi, \qquad (4.62)$$

while the scalar field equation (2.7) is modified to

$$\Box\varphi = \zeta^2(T - 4\Lambda). \qquad (4.63)$$

The way the new term -4Λ enters can be understood by noting that Λ appears in (4.62) in a combination $T_{\mu\nu} - g_{\mu\nu}\Lambda$.

Again in the spatially uniform and flat, isotropic Robertson–Walker universe, we may also expect that there is a solution of the scalar field that is spatially uniform, depending only on t, as will be verified shortly. Equations (4.61)–(4.63) are now cast into the forms

$$6\varphi H^2 = \tfrac{1}{2}\epsilon\dot{\phi}^2 + \Lambda + \rho - 6H\dot{\varphi}, \qquad (4.64)$$
$$\ddot{\varphi} + 3H\dot{\varphi} = 4\zeta^2\Lambda, \qquad (4.65)$$
$$\dot{\rho} + 4H\rho = 0, \qquad (4.66)$$

where we assumed radiation-dominance for simplicity, for the moment.

We easily find that these equations allow a surprisingly simple solution. Put

$$H = 0. \qquad (4.67)$$

From (4.66) we then derive

$$\rho = \text{constant.} \qquad (4.68)$$

Further substituting (4.67) into (4.65) yields

$$\ddot{\varphi} = 4\zeta^2\Lambda, \qquad (4.69)$$

which solves to give

$$\varphi(t) = 2\zeta^2\Lambda t^2 + \varphi_1 t + \varphi_0, \qquad (4.70)$$

where φ_1 and φ_0 are constants. For $t \to \infty$ we obtain an asymptotic solution

$$\phi \approx \sqrt{4\Lambda\zeta^2\xi^{-1}}\,t = \sqrt{\frac{4\Lambda}{6\xi + \epsilon}}\,t, \qquad (4.71)$$

where we recalled that $\varphi = (\xi/2)\phi^2$.

By using these results in (4.64), we determine the constant in (4.68), finding

$$\rho = -3\Lambda\frac{2\xi + \epsilon}{6\xi + \epsilon}. \qquad (4.72)$$

The same solution as that given by (4.67) and (4.72) was obtained by Wetterich [71], though we draw a different conclusion.

From (4.72) we immediately arrive at two conclusions. First the fact that ρ is proportional to Λ shows that we are not able to reach a sensible theory without Λ by the limiting procedure $\Lambda \to 0$ in our present model. Also from (4.72) we find that the physical condition $\rho > 0$ is satisfied only if

$$\epsilon = -1 \tag{4.73}$$

and

$$2 < \xi^{-1} < 6 \tag{4.74}$$

are obeyed simultaneously. The relation (4.38) is then translated into

$$\zeta^{-2} = 6 - \xi^{-1} < 4. \tag{4.75}$$

One may also express (4.72) as

$$\rho/\Lambda = 12\left(\zeta^2 - \tfrac{1}{4}\right), \tag{4.76}$$

which it might be convenient to use later.

From (4.64) with $H = 0$, we find that $\Lambda > 0$ and $\rho > 0$ are realized only with (4.73); a positive Λ is canceled out by the negative scalar-field energy.

In passing we describe briefly how the results differ if we assume the existence of dust matter. We still find solutions with (4.67). The right-hand sides of (4.65) and (4.69) are replaced by $\zeta^2(4\Lambda + \rho)$. In place of (4.72) we then obtain

$$\rho = -\frac{2}{3}\Lambda\frac{2\xi + \epsilon}{4\xi + \epsilon}, \tag{4.77}$$

leaving the conclusion (4.73) unchanged, while replacing (4.74) by $2 < \xi^{-1} < 4$.

In either case, (4.67) shows that the universe is *static*. Also, from (4.72) or (4.77), we have ρ as large as Λ. These conditions are far from reality.

Nevertheless, the asymptotic absence of an exponential growth of the scale factor might be viewed as a successful taming of the cosmological constant, albeit coming with a highly unrealistic situation. The same conclusion was reached also by Dolgov [72] and Ford [73], who restricted their consideration to $\rho = 0$. In this "vacuum solution," the condition (4.73) has never been recognized. Also they failed to derive solutions in the E frame, as we will show soon. This tends to invite the criticism that the result does not make sense because of the unrealistically rapid diminution of the effective gravitational "constant," $G_{\text{eff}} \sim \phi^{-2} \sim t^{-2}$, expected from (4.71).

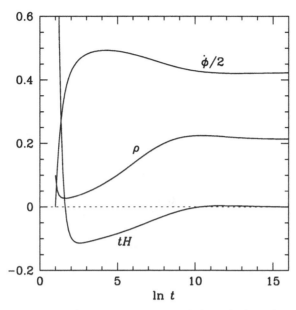

Fig. 4.1. An example of the solutions of (4.64)–(4.66) showing how H, ρ, and $\dot{\phi}$ approach the asymptotic behaviors (4.67), (4.68), and (4.71), respectively. The initial values at $\ln t_1 = 1$ are $\phi_1 = 0.25$, $\dot{\phi}_1 = 0$, and $\rho_1 = 0.1$ [70].

As we will see shortly, however, the same static universe in the J frame looks as if it is expanding in the E frame. Before going into details on this point, however, we are going to discuss whether the solution given by (4.67)–(4.72) is stable. In fact the solution is an "attractor," to which solutions of any initial values tend. An example of numerically integrated solutions is shown in Fig. 4.1, in which we find how each of the curves approaches asymptotically the values given by (4.67)–(4.72). For example, $a(t)$ starts varying (hence we have nonzero $H(t)$ at early times), but gradually approaches a constant (hence we have $H = 0$).

One may also notice the fact that H enters the equations (4.64)–(4.66) without a derivative. We therefore eliminate H algebraically, and study how $\varphi, \dot{\varphi}$, and ρ behave in three-dimensional phase space. Figure 4.2 shows a two-dimensional cross section in the plane of $\dot{\phi}^2 = \dot{\varphi}^2/(2\xi\varphi)$ and ρ. The four solutions shown correspond to different initial values, starting from different positions, finally converging spirally toward the common point. This is what is meant by an attractor.

4.4.2 Solution in the E frame

We now consider the same model as above in the E frame [69, 70, 74]. We apply the conformal transformation (3.6), (3.38), and (3.40) to (4.60).

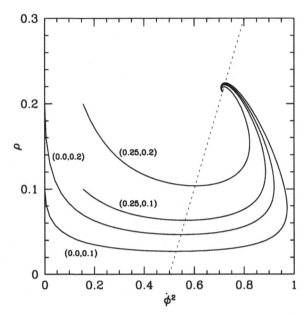

Fig. 4.2. A two-dimensional cross section in the plane of $\dot{\phi}^2$ and ρ, cut from the three-dimensional phase space of $\varphi, \dot{\varphi}$, and ρ. Four solutions converge to a common point. With the common initial value $\phi_1 = 0.25$, other initial values of $\dot{\phi}_1^2$ and ρ_1 are shown in the parentheses [70].

Terms except for Λ were already obtained in (3.34). The remaining term is

$$-\sqrt{-g}\Lambda = -\sqrt{-g_*}\Omega^{-4}\Lambda, \tag{4.78}$$

thanks to (3.8). On combining these we reach

$$\mathcal{L}_{\mathrm{BD\Lambda}} = \sqrt{-g_*}\left(\tfrac{1}{2}R_* - \tfrac{1}{2}g_*^{\mu\nu}\,\partial_\mu\sigma\,\partial_\nu\sigma - \Lambda e^{-4\zeta\sigma} + L_{*\mathrm{matter}}\right), \tag{4.79}$$

which will be the starting point of this subsection. We should notice that Λ is now multiplied by $e^{-4\zeta\sigma}$. In other words, the Λ-term acts as a potential of the field σ in the E frame. It is the presence of this potential that causes major differences between the models with and without Λ, as we will see.

In the initial J frame, the Λ-term looks as if it has no relation to the motion of the scalar field. It does exert a force on σ, however, in the E frame, in which there is nothing ambiguous because σ and the metric are diagonalized keeping each part independent of the other, as was explained in Chapter 2. In the J frame, in contrast, the same interaction affects the scalar field *indirectly* through the mixing interaction. For this reason, one might be mistaken if one were to look only at the interaction terms, ignoring the kinetic term.

The potential term for σ obtained in this way is *exponential*. The scalar field keeps moving toward infinity, rolling down the slope steadily. The slope becomes milder as it proceeds, with the "speed" $\dot\sigma$ becoming ever slower.

In this way the total energy (kinetic plus potential) of σ reduces gradually. Since this energy is directly connected to Λ, we find that the effect of the cosmological constant becomes smaller and smaller with time. This is nothing but the mechanism of making Λ "harmless," as we saw in the previous section. In the present method, moreover, the effect of Λ decreases as t_*^{-2}, allowing us to understand the smallness of 120 orders of magnitude naturally. In order to formulate these results in a more accurate manner, we try to solve the cosmological equations in the E frame.

We begin by writing down "Einstein's equation" and the scalar-field equation derived from (4.79):

$$G_{*\mu\nu} = T^{(\sigma)}_{*\mu\nu} + T_{*\mu\nu} = T_{\mu\nu}, \tag{4.80}$$

$$\Box_*\sigma + 4\zeta\Lambda e^{-4\zeta\sigma} = \zeta T_*. \tag{4.81}$$

Note that we included the potential term $\Lambda e^{-4\zeta\sigma}$ in (4.80), with its far-right-hand side defining a total energy–momentum tensor $T_{\mu\nu}$ in the E frame. Also, the right-hand side of (4.81) comes from (3.54), or from

$$-\frac{\partial L_{*m}}{\partial\sigma} = \zeta T_*. \tag{4.82}$$

Recall that $T_{*\mu\nu}$ for the matter now depends on σ in the E frame.

Obviously, the matter energy–momentum tensor fails to satisfy the covariant conservation law, unlike in the J frame. The revised law is given by

$$\nabla_\mu T_*^{\mu\nu} = g^{\nu\mu}(\partial_\mu\sigma)\zeta T_*, \tag{4.83}$$

which comes from the Bianchi identity, or the covariant conservation law

$$\nabla_\mu T^{\mu\nu} = 0. \tag{4.84}$$

A simplified derivation of (4.83) will be presented in Appendix J.

In comparing these with equations in the J frame, we find a similarity, though they are not exactly the same, between (4.63) and (4.81), on the one hand, and the difference between the right-hand sides of (4.61) and (4.83), on the other.

We now move to consider the Robertson–Walker metric (4.3) in the form

$$ds^2 = -dt^2 + a^2(t)\,d\vec{x}^2, \tag{4.85}$$

to which we apply the conformal transformation (3.2):

$$ds_*^2 = \Omega^2\,ds^2 = -\Omega^2\,dt^2 + \Omega^2 a^2(t)\,d\vec{x}^2. \tag{4.86}$$

We write this in the form

$$ds_*^2 = -dt_*^2 + a_*^2(t_*)\, d\vec{x}^2, \qquad (4.87)$$

by which we *define* t_* as the cosmic time in the E frame, and $a_*(t_*)$ as the scale factor. We then have

$$dt_* = \Omega\, dt, \qquad (4.88)$$

$$a_* = \Omega a. \qquad (4.89)$$

One may wonder why we have a coordinate transformation (4.88), although a conformal transformation itself has nothing to do with the former. The answer lies in the fact that the cosmic time is defined in any reference frame always with the condition $g_{00} = -1$; it should be the proper time in any of the freely falling systems in the universe.

In this connection one may also ask whether $d\vec{x}$ should be transformed when we derive (4.89). Notice, however, that $d\vec{x}$ is a *co-moving coordinate*, which involves a set of well-defined angles not subject to continuous change of scale any longer. With these remarks in mind, we assume a Robertson–Walker metric of the same "form."

Now that the metric has been determined, we try to re-write (4.80) and (4.81). We again emphasize that we seek a solution of σ that depends only on t_*.

"Einstein's equation" reads

$$3H_*^2 = \rho_\sigma + \rho_*, \qquad (4.90)$$

where

$$\rho_\sigma = \tfrac{1}{2}\dot{\sigma}^2 + V(\sigma), \qquad (4.91)$$

$$V(\sigma) = \Lambda e^{-4\zeta\sigma}. \qquad (4.92)$$

We choose the traceless condition, $T_* = 0$, corresponding to radiation-dominance. The σ-equation is therefore given by

$$\ddot{\sigma} + 3H_*\dot{\sigma} + V'(\sigma) = 0. \qquad (4.93)$$

We first try to find the solution corresponding to the attractor asymptotic solution (4.67), (4.68), and (4.71) in the J frame. For this purpose we start by re-writing (4.71) as

$$\phi = 2\sqrt{\Lambda\xi^{-1}}\zeta t. \qquad (4.94)$$

We reproduce (3.38) and (3.40):

$$\phi = \xi^{-1/2}e^{\zeta\sigma}, \qquad (4.95)$$

$$\Omega = e^{\zeta\sigma}. \qquad (4.96)$$

Now, on substituting (4.94) with (4.95) combined with (4.96) into (4.88), we find

$$dt_* = t\, dt, \tag{4.97}$$

and hence

$$t_* = t^2, \tag{4.98}$$

where we ignored the multiplicative factor which turns out to be irrelevant. By applying (4.98) and (4.94) to (4.96) in (4.89), we find that $a = $ constant, as implied by (4.67), is translated into

$$a_* = t_*^{1/2}, \quad \text{or} \quad H_* = \tfrac{1}{2}t_*^{-1}. \tag{4.99}$$

Further using (4.97) in (4.94) and (4.95), we also find

$$e^{\zeta\sigma} = e^{\zeta\bar{\sigma}}t_*^{1/2}, \tag{4.100}$$

where $\bar{\sigma}$ is an arbitrary constant to be determined. From (4.100) we have

$$V = \Omega^{-4}\Lambda = \Lambda e^{-4\zeta\bar{\sigma}}t_*^{-2}, \tag{4.101}$$
$$\zeta\dot{\sigma} = \tfrac{1}{2}t_*^{-1}, \tag{4.102}$$
$$\zeta\ddot{\sigma} = -\tfrac{1}{2}t_*^{-2}, \tag{4.103}$$

where an overdot implies differentiation with respect to t_*. Substituting these into (4.93) yields

$$\Lambda e^{-4\zeta\bar{\sigma}} = \frac{1}{16}\zeta^{-2}, \tag{4.104}$$

which we can use as a constraint to determine $\bar{\sigma}$. By using this in (4.101) we obtain

$$V = \frac{1}{16}\zeta^{-2}t_*^{-2}. \tag{4.105}$$

Finally, using (4.99), (4.102), and (4.105) in (4.90), we determine

$$\rho_\sigma = \frac{3}{16}\zeta^{-2}t_*^{-2}, \tag{4.106}$$
$$\rho_* = \tfrac{3}{4}\left(1 - \tfrac{1}{4}\zeta^{-2}\right)t_*^{-2}. \tag{4.107}$$

Also we may re-write (4.100) as

$$\sigma(t_*) = \bar{\sigma} + \tfrac{1}{2}\zeta^{-1}\ln t_*. \tag{4.108}$$

In this way we obtained a set of asymptotic solutions (4.99) and (4.106)–(4.108) as an E frame counterpart of (4.67), (4.68), and (4.71). It should also be pointed out that it is even simpler to derive these results without recourse to the J-frame solution. We simply assume that all the terms on both sides of (4.90) behave as t_*^{-2}. In fact (4.108) is nothing but the condition that $e^{-4\zeta\sigma}$ in $V(\sigma)$ falls off like t_*^{-2}.

We notice that ρ_σ given by (4.106) is derived originally from Λ, and may be called an "effective" cosmological constant in the E frame:

$$\Lambda_{\text{eff}} = \rho_\sigma, \tag{4.109}$$

which decreases like $\sim t_*^{-2}$, probably justifying our calling it a *decaying cosmological constant*. This seems to be part of the "dark energy" in cosmology.

It might be interesting to use (4.106) and (4.107) combined with (4.109) to derive the relation

$$\rho_*/\Lambda_{\text{eff}} = 4\left(\zeta^2 - \tfrac{1}{4}\right), \tag{4.110}$$

which is to be compared with (4.76) for a possible counterpart in the J frame. The difference by a factor of three would have been absent were σ unchanging. This is another example of similar differences between the left-hand side of (4.62) and that of (4.80), in the J frame and the E frame, respectively. Both are the 00-components of Einstein's tensor, which are to be related to each other by Ω^2, but their contents in terms of other types of tensors as shown on the respective right-hand sides appear differently, insofar as the contributions from the variable part of the scalar field are concerned. This is simply a manifestation of the fact that the theory is *neither* invariant *nor* covariant under conformal transformations.

In this way we came to understand the "first face" of the problem of the cosmological constant. There seems to be, however, a "second face," namely the indicated presence of a nonzero value instead of a mere upper bound. The above result is still insufficient in this respect. We postpone this issue until later sections.

We only point out here that t_*^{-2} represents the magnitude of the critical density. If the universe expands, in particular according to $a(t) \sim t_*^\alpha$, we have

$$\rho_{\text{cr}} = 3H_*^2 = 3\alpha^2 t_*^{-2}. \tag{4.111}$$

The above argument tells us that Λ_{eff} behaves in the same manner as the ordinary density, which decreases as $\sim t_*^{-2}$ as long as the universe expands according to a "power law," the behavior later called "scaling."

As another point to be noticed, we find, from (4.107),

$$\zeta > \tfrac{1}{2}, \tag{4.112}$$

for the condition $\rho_* > 0$. This is obviously related to the fact that too small a ζ implies too rapid a variation of $\sigma(t_*)$, according to (4.108), and thence too large a ρ_σ, which can be consistent with (4.90) only at the expense of a positivity of ρ_*.

A similar analysis for the case of dust-dominance is found to replace (4.112) by

$$\zeta > \sqrt{3}/4, \tag{4.113}$$

which is not very much different from $\frac{1}{2}$.

We show, however, that we have another attractor solution, the σ-dominant solution [74]. For this purpose we demand that $\rho_* = 0$, considering the equation

$$3H^2 = \rho_\sigma, \tag{4.114}$$

together with (4.93). In the same way as before, we obtain (4.108) and

$$a_* = t_*^p, \tag{4.115}$$

where

$$p = \tfrac{1}{8}\zeta^{-2}. \tag{4.116}$$

The possible significance of this solution will be discussed later.

Taking aside this "vacuum solution," we find that the condition (4.112) has another important consequence; this model *does not allow* the general-relativity limit $\zeta \to 0$ (plus Λ). This may be understood by observing that in the latter limit the potential term (4.92) goes back to a pure constant Λ, hence leaving no room for the power-law expansion of the universe.

Incidentally, we note that (4.112) implies that

$$\xi^{-1} > 2, \tag{4.117}$$

which agrees with one of the conditions (4.74) obtained in the J frame.

We conclude that the universe *looks quite different* depending on which conformal frame we use to describe it. There is no expansion in the J frame, whereas there is an expansion in the E frame in accord with the usual expansion of radiation-dominance. Which is the correct description from a physical point of view?

Before we propose a possible remedy, we add some other aspects of the current model. We first discuss another example of a conflict with the standard cosmology. In our calculation for the dust-dominated universe, we are going to predict the *same* expansion law as that for the radiation-dominated era. The key lies in (4.108), which assures that $V(\sigma)$ behaves like t_*^{-2} in the same way as other terms on the right-hand side of (4.80), and holds true irrespective of whether the universe is dominated by radiation or dust. As a consequence of the right-hand side of (4.83), the dust density ρ_* obeys the equation

$$\dot{\rho}_* + 3H_*\rho_* = -\dot{\sigma}\zeta\rho_*. \tag{4.118}$$

Assume that

$$\rho_* = \rho_0 t_*^{-2}, \tag{4.119}$$

as in (4.107), and

$$H_* = \alpha t_*^{-1}. \tag{4.120}$$

Substituting them into (4.118) results in $\alpha = \frac{1}{2}$ instead of $\frac{2}{3}$. The right-hand side of (4.118) turns out to act in the same way as the pressure term of the radiation fields. We find in this way that the difficulty in the dust-dominated era is rooted deep in the foundation of the model.

In addition to the above asymptotic solution, we may have a solution presumably corresponding to primordial inflation. Suppose that, at the initial times, at the very beginning of classical dynamics, we may consider $\sigma, \dot{\sigma}$, and ρ negligibly small compared with Λ. Then (4.90) will go back to (4.6) with ρ ignored. We have thus (4.15), an inflation, which will last until some of the conditions stated above fail to be satisfied. This is the reason why we expected that our gravitational scalar field can be viewed as playing the role of "inflaton" for some duration of the time. One might even expect that a sufficient amount of energy would have been created even if the potential falls off smoothly, hence without an oscillation of the scalar field [74].

We point out, however, that the above conclusion is still far from assuring that the E-frame description is better than the J-frame description. In the rest of this subsection we show that a drawback of the E frame is that it produces too much time-variation of particle masses, thus suggesting a possible remedy.

By combining (3.63) with (4.100) we find that the mass in the E frame decreases with time as

$$m_* \sim t_*^{-1/2}. \tag{4.121}$$

This much time-dependence is not tolerated. During the period of ~100–1000 s (the first 3 min) for the primordial nucleosynthesis, the neutron mass, for example, should have reduced by as much as $1 - 1/\sqrt{10} \approx 70\%$, in spite of the fact that theoretical analyses are based on quantum mechanics in which particle masses are kept strictly constant. It should be emphasized that this is independent of the possible question of whether someone had prepared a measuring device, an atomic clock, for example, to check the time-dependence of mass. If mass does vary with time, then quantum mechanics as a tool for studying the past formation of light elements has to be revised accordingly, though that is far from reality.

In passing we may offer an explanation for how a static universe in the J frame may be expected from the varying mass in the E frame. The only constant dimensional number in the E frame is a gravitational constant. Masses are measured as varying in terms of units of this constant. We know, on the other hand, that, in the radiation-dominated epochs, the expanding behavior of the universe has in principle nothing to do with how the masses of low-energy particles vary with time. The law $a_* \sim t_*^{1/2}$ is determined completely by the conservation equation of matter

$$\dot{\rho}_* + 4H_*\rho_* = 0, \tag{4.122}$$

to which nonrelativistic particles contribute only a little. In this context the only way expansion of the universe affects physics is through the temperature falling off like $T \sim a_*^{-1} \sim t_*^{-1/2}$. The temperature then gives an average energy of the particles measured in units of the masses. In this sense the expansion is measured in units of masses.

Now accept that masses vary as $m_* \sim t_*^{-1/2}$. We also assume that a length scale, or, according to special relativity, a time unit, is provided only by particle masses through the relation $\tau \sim m_*^{-1} \sim t_*^{1/2}$. As it happens, this "microscopic time unit" varies at the same rate as that at which a_* expands. Measuring the expansion of the universe with a meter-stick that changes at the same rate obviously fails to detect expansion.

One may also define a time variable $\tilde{t}$ in terms of the "microscopic" time t_* by

$$d\tilde{t} = \frac{dt_*}{\tau} \sim t_*^{-1/2} \, dt_*. \tag{4.123}$$

Integration gives

$$\tilde{t} \sim t_*^{1/2}, \tag{4.124}$$

which, on comparison with (4.98), shows that $\tilde{t} = t$, namely we are going back to the J frame.

We then face a dilemma; should we choose the J frame because there is a time-independent particle mass, or the E frame because of the expanding universe? One might further ask whether this applies only to the prototype BD model in the strict sense. Can one recover a more or less standard conclusion by modifying part of the prototype model?

In order to answer the last question at least partially, we studied the model in which the Λ-term is not completely constant but is multiplied by a power of the scalar field [70], as had been suggested by Wagoner [9] and also by Endō and Fukui [76]:

$$\Lambda \to \Lambda\phi^q, \qquad (4.125)$$

where q need be neither an integer nor positive. This modification seems likely from the KK approach in D dimensions. In Appendix A, (A.23) shows in fact that q is given by $2 - 4/(D - 4)$. Generally speaking, a four-dimensional Lagrangian may assume the form of a monomial of ϕ. Detailed analysis in Appendix K shows that the conclusion from $q = 0$ hardly seems noticeably different unless $|q|$ differs from 0 considerably, to an extent that appears unacceptable. In short, any minor change from the prototype BD model is unlikely to be a good approximation to the standard cosmology. In other words, a standard picture is so different from what we conclude from the prototype model that the necessary change, if any, should be made at a far more fundamental level.

It seems appropriate here to include an argument on the contribution from the vacuum energy of the matter fields, which we previously mentioned as the second origin of what is known as a cosmological constant. The vacuum energy is obviously of quantum nature, yet appears to be about 60 orders of magnitude too large compared with the critical density.

We admit that the problem is formidable. We may derive sensible results only by assuming that we can overcome this issue somehow, in a way that will still leave some questions unanswered. Nevertheless, we outline here an indication that we may effectively ignore this effect entirely.

We emphasize that this is an issue that can arise even without the primordial cosmological constant, the first origin. Consider the standard model with neither Λ nor the scalar field, and imagine that the vacuum energy, usually chosen to be $M_{\rm ssb}^4$ with $M_{\rm ssb}$ for the mass scale of violation of supersymmetry, is not present from the outset, but *accumulates* gradually, starting from an infinitesimal amount. We assume that this process takes place in a spatially uniform manner, no matter how rapidly or slowly it may proceed. Also, at the "initial" time of the process we assume that the universe is expanding with the density of ordinary matter falling off like t_*^{-2}. A similar situation can be seen if a mini-inflation occurs in the model of two scalar fields, as will be discussed in Chapter 5. We require that the same is the case even in the absence of the scalar fields.

Since the vacuum-energy contribution $\rho_{\rm vac}$ arises due to the nongravitational interactions among matter fields, it is part of ρ_*, which is the sum with the usual nonvacuum energy $\rho_{\rm nvc}$:

$$\rho_* = \rho_{\rm vac} + \rho_{\rm nvc}. \qquad (4.126)$$

Let us consider first a radiation-dominated era in which the traceless condition holds for the vacuum energy as well. We then have the energy density that obeys

$$\dot{\rho}_* + 4H_*\rho_* = 0, \qquad (4.127)$$

which implies that ρ_* decreases in the expanding universe, namely $H_* > 0$. From this consideration, the vacuum energy is created only at the expense of the nonvacuum energy. This ρ_* is substituted into the right-hand side of Einstein's equation, from which it follows that H_* decreases with time; hence we have a decelerating universe, as usual.

In this approach we obviously find that ρ_{vac} has no chance of becoming so large that the nature of expansion changes drastically from a power law, as long as the physical condition $\rho_{\mathrm{nvc}} > 0$ is maintained. There must be a built-in mechanism to suppress the building up of the vacuum energy if it occurs in an expanding universe.

One may compare this with the situation in which a scalar field plays an important role. The covariant conservation law (4.127) still holds, as will be confirmed from (4.83) for $T_* = 0$, but another contribution, namely ρ_σ from the scalar field, is present on the right-hand side of Einstein's equation, implying that the universe is free to expand without being constrained by the law of conservation of matter. This makes a difference from the model without a gravitational scalar field.

In this context we may thus expect that the vacuum energy can never make a major contribution to the whole matter energy density. We may reasonably extend the argument to the dust-dominated era, in which the pressureless condition need not apply to the vacuum part.

Though the argument is only plausible, and admittedly still far from being complete, we proceed further to offer a conjecture that the above-mentioned built-in mechanism remains at work even in the presence of scalar fields, hoping that the mechanism itself should apply to nongravitational processes. In this way we can find a reason for assuming that the quantum vacuum energy remains insignificant. Without further elaboration we simply ignore the contribution.

4.4.3 A proposed revision and remarks

We consider that the dilemma we mentioned above is truly serious and try to resolve it. In this connection we remind the reader that the law of cosmological expansion in the E frame in the radiation-dominated era is determined solely on the basis of Einstein's equation and the conservation law (4.127), being independent of whether masses of low-energy particles vary with time. Taking advantage of this fact, we hope to reach our goal by *favoring the E frame*, reconsidering the way masses vary as a separate issue [69, 70].

Consider a simplified situation in which matter is represented by a real scalar field Φ. The idea might be implemented by the *assumed* E-frame mass term of the form

$$\mathcal{L}_{\text{mass}} = -\tfrac{1}{2}\sqrt{-g_*}m_\Phi^2\Phi_*^2, \qquad (4.128)$$

instead of the last term on the right-hand side of (3.60), where m_Φ is meant to be a constant. This would make it possible for neutrons, for example, to keep their masses constant, allowing us a consistent understanding of the primordial nucleosynthesis of light elements, in accordance with the conventional use of quantum mechanics.

The coupling (4.128) may be transformed back to the J frame. Consider an interaction term in the J frame given by

$$\mathcal{L}_{\text{mass}} = -\tfrac{1}{2}\sqrt{-g}f_\Phi^2\phi^2\Phi^2, \qquad (4.129)$$

where f_Φ is a *dimensionless* coupling constant. By applying (3.56) and (3.59), we easily re-express (4.129), obtaining the E-frame Lagrangian:

$$\mathcal{L}_{\text{mass}} = -\tfrac{1}{2}\sqrt{-g_*}f_\Phi^2\Omega^{-2}\phi^2\Phi_*^2. \qquad (4.130)$$

Using (3.57) allows us to identify the result with (4.128) by imposing the relation

$$m_\Phi = \xi^{-1/2}f_\Phi. \qquad (4.131)$$

In our unit system we have $m_\Phi \sim 1\,\text{GeV} \sim 10^{-18}$, for example, and hence $f_\Phi \sim 10^{-18}$, as long as ξ is of the order of unity. The smallness of f_Φ of this kind had driven Dirac to propose his hypothesis of varying G. According to Landau [51] and others [52], however, let us put

$$10^{-18} = e^{-\beta}, \qquad (4.132)$$

finding $\beta \sim 18 \times 2.30 \approx 41$, which is rather close to $\alpha^{-1} \approx 137$, the inverse of the fine-structure constant, as was discussed in subsection 1.3.4. This highly nonlinear relation to coupling constants of nongravitational interactions appears to provide us with an alternative view of very small or large numbers in nature. We here acknowledge a small number f_Φ as it is.

In (4.130) and (4.131), we realize that the scalar field does not appear in the E frame, insofar as the world of matter is concerned. There is the space-time-dependent σ field still in the E frame, but this never shows up as something that mediates a force between ordinary matter objects. It still couples to matter through the usual minimal coupling of the metric field, and hence will act as cosmological "dark matter," without having any explicit coupling to matter. Now the WEP or its violation is expected to be observed through the forces among matter objects. In this sense, we find that no process violating the WEP can possibly arise from the exchange of σ.

The assumed *presence* of ϕ in the J-frame matter Lagrangian, indicating violation of the WEP in general as was discussed previously, turns out to be "harmless." This "magical" mechanism remains at work for a fermion matter field as well, if we follow a special but simple rule that the coupling to matter observes dilatation symmetry in four dimensions, i.e. with a dimensionless coupling constant.

As an added advantage we point out that we now obtain the "correct" expansion law $a_* \sim t_*^{2/3}$ for dust-dominance. Recall that the prototype BD model with simply added Λ entails $a_* \sim t_*^{1/2}$, as we remarked around (4.118)–(4.120) before. Since the scalar field no longer couples to matter directly in the revised model, we have no term on the right-hand side of (4.118), hence allowing the behavior of the scale factor to be determined as usual.

In this way we arrived at a "favored" outcome; even in the presence of a primordial cosmological constant, we recover the standard scheme for an expanding universe, the scalar field acting only as dark matter without offending the WEP too much for low-energy phenomena. The parameter ζ is determined by (4.110), the ratio of the ordinary matter to the effective Λ which is supposed to be a remnant of the decaying cosmological constant. Solar-system experiments and analyses of the primordial nucleosynthesis no longer constrain either ζ or ω.

Another further advantage to be emphasized is that the mass dimension which produces particle mass through (4.131) comes essentially from M_P that had been introduced to give G. Mass generation of this kind can tacitly be interpreted as providing a common origin for the Planck mass and particle masses [68]. This might be restated also as a common origin for M_P and the vacuum expectation value, which plays a crucial role in the electroweak and GUT unification models.

We admit, however, that there are some questions about the relevance of the proposed model. They will be summarized below.

First we point out that the four-dimensional scale invariance which is expected to constrain the way in which coupling to matter enters does not seem to be supported by any models of fundamental theories, such as string theory, though this is an area where detailed and unique predictions of string theory, for example, are still most sparse. At this moment we mean only that coupling to matter of the scalar field as forbidden by Brans and Dicke may provide a potential remedy for theoretical models that incorporate Λ. On the other hand, scale invariance is appealing [68], because dimensionless coupling constants in four dimensions mean that there is a boundary between non- and super-renormalizable theories.

A closely related, and probably more serious, issue is that of how the above-mentioned *classical* effects are affected by quantum theory. In the next chapter we are going to discuss the quantum *anomaly* which is known to alter some of the invariance principles established at a classical level. We specifically consider loop effects coming from nongravitational interactions among matter fields. Two of the obvious effects are a spontaneous breakdown of dilatation invariance, thus causing the massless ϕ as a NG boson to acquire a finite mass, on the one hand, and bringing the scalar field into coupling directly to ordinary matter fields, on the other. This will give the scalar field the nature of the non-Newtonian field, which is characterized by a finite force-range and violation of the WEP. This part of the analysis which will be presented in more detail is somewhat complicated and we are still short of drawing a final conclusion. For this reason, and also recalling that the anomaly arises because the invariance under scale transformation is lost for dimensionality off the value 4, we pursue in this chapter another theoretical possibility, namely that scale invariance holds for *any* dimensionality.

In this approach we replace (4.129) by

$$\mathcal{L}_{\text{mass}} = -\tfrac{1}{2}\sqrt{-g}\,f_\Phi^2\phi^{2/(d-1)}\Phi^2, \tag{4.133}$$

where $d = D/2$ [70]. We assumed naturally that the matter field Φ enters here only quadratically. Noting that the mass dimensions of ϕ and Φ, determined from the kinetic terms, are both $d-1$, we easily find that the coupling constant f_Φ is always dimensionless.

The relations (3.57) and (3.59) are replaced by

$$\Omega = \xi^{1/(D-2)}\phi^{1/(d-1)} \tag{4.134}$$

and

$$\Phi = \Omega^{d-1}\Phi_*, \tag{4.135}$$

respectively. From these relations we confirm that (4.133) in the J frame is indeed re-expressed as (4.128) if (4.131) is replaced by

$$m_\Phi = \xi^{-1/(D-2)}f_\Phi. \tag{4.136}$$

In contrast to the model of (4.129) for which scale invariance is correct only in four dimensions, and hence the quantum effects imply the coupling to matter of σ in the E frame, the present model is found to be immune against quantum effects. One might complain that the coupling (4.133) is too artificial. For the sake of simplicity, and also because quantum corrections are likely to be rather small, as will be shown in Chapter 5, we will analyze the classical model first. The model corrected for the quantum effects will be discussed in Chapter 6.

5

Models of an accelerating universe

We now move on to discuss the second face of the problem of the cosmological constant, which was highlighted recently by the discovery of the acceleration of the universe. This chapter will first review briefly how searching for "dark energy" has come finally to a spatially flat universe well described by a cosmological constant Λ of a size smaller than but nearly comparable to the critical density. For a number of reasons, we consider that this Λ is not a true constant but is mimicked most naturally by a scalar field.

In section 5.1, we sketch what the development has been like mainly on the observational front, culminating in the conclusion that we have an accelerating universe.

As a possible theoretical model discussed recently, we first review in section 5.2 the results of "quintessence," a name mainly indicating a cosmological scalar field. Since this is a phenomenological approach that is not necessarily constrained rigorously by the scalar–tensor theory, our focus is mainly on the assumed inverse-power potential. A primary concern is the question of how naturally the initial conditions for the scalar field can be chosen. A relevant question is that of whether the scalar-field energy falls off in the same way as the ordinary matter density ("scaling"), or approaches the latter starting from different values ("tracking").

Section 5.3 describes how the same analysis is extended to the assumed brane structure for the underlying cosmology. After examining the inverse-power potential extensively, we reach the conclusion that the presence of a brane helps to widen the range of allowed initial values of the quintessence field. It appears that the inverse-power potential fits in better with the idea of the brane cosmology which offers a novel approach to understanding the hierarchy problem in unified theories.

We then come back in section 5.4 to the scalar–tensor theory, describing what modification has to be made in order for it to correspond to the observed accelerating universe. We decided to introduce another scalar field. Though this may sound a little artificial, it shows unambiguously how we can reproduce the recent observational results in accordance with other aspects of standard cosmology, including a successful understanding of nucleosynthesis, among other things.

According to this approach the behavior of the extra acceleration turns out to be just one of repeated occurrences during the whole of cosmological evolution. From this point of view, an acceleration observed during the present epoch should not last forever; it is just a "mini-inflation" that will soon die out, with the universe resuming an ordinary power-law expansion. This seems to suggest a new perspective on the entire evolution of the universe, particularly in connection with the question, often called the "coincidence problem," of how likely it is that we are now witnessing the occurrence of a mini-inflation.

As one of the outcomes, we may depart from the common-sense attitude that we are in an attractor solution in the asymptotic phase which we are supposed to arrive at for whatever initial values. Instead we might be still in a transient state. This may have something to do with the partially chaotic nature of the solution.

5.1 Dark energy

Around the early 1990s, a new question started to emerge, bothering cosmologists, as the Hubble Space Telescope (HST) sent back its first pictures of a small corner of the northern sky showing us startling images of objects as dim as 30th magnitude. It accelerated activity from the ground as well. Observations of various phenomena started to point gradually to the existence of a cosmological constant, which is small in the sense that it stayed below the hitherto-obtained upper bound, the critical density. It was, in retrospect, the beginning of the second phase in the development of the contemporary version of the problem of the cosmological constant.

Following the custom, we introduce a parameter Ω_Λ defined by

$$\Omega_\Lambda = \Lambda/\rho_{\mathrm{cr}}, \tag{5.1}$$

where

$$\rho_{\mathrm{cr}} = 3H_0^2, \tag{5.2}$$

given in terms of the Hubble constant $H_0 = \dot{a}/a$ estimated for the present epoch. We also use the notation h_0 expressing H_0 in units of $100\,\mathrm{km\,s^{-1}\,Mpc^{-1}}$.

In addition to Ω_Λ, other kinds of ratios, often called density parameters, are used conveniently:

$$\Omega_m = \frac{\rho_m}{\rho_{cr}}, \quad \Omega_b = \frac{\rho_b}{\rho_{cr}}, \quad \Omega_k = \frac{\rho_k}{\rho_{cr}}, \tag{5.3}$$

for the matter energy density, baryon density, and curvature density, $\rho_k = -(\frac{1}{3})ka^{-2}$, respectively. Since Ω_b and Ω_k are now considered small, most of the matter density should come from dark matter, which is supposed to be clumpy, being different from dark energy, as we are shortly going to mention.

The HST measurement of Virgo Cepheids began to tell us that the value of the Hubble constant is as high as $h_0 \sim 0.8$, in contrast to the result $h_0 \sim 0.5$, which had steadily been being produced before from measurements for type-Ia supernovae. One of the most serious problems arising from a high H_0 was that it would predict that the age of the universe is as little as $(7-8) \times 10^9$ years if 3-space is flat, considerably younger than the oldest globular clusters, which are believed to be as old as $\sim 1.4 \times 10^{10}$ years.

An easy solution to this "age problem" was to call for Λ. The required size for this purpose is of the order of unity in terms of Ω_Λ. This might have been a renaissance of solutions of "newly found" observations that had been repeated before many times. We recall that Eddington once proposed the same thing when h_0 was believed to be 5, predicting the age $\sim 2 \times 10^9$ years.

This time, however, the argument for Λ was gathering momentum from other sources as well [77], including the following.

- The large-scale structure of galaxies. Eftathiou *et al.* obtained the observational result that distribution of galaxies on scales larger than 10 Mpc is less smooth than that predicted by the cold-dark-matter (CDM) scenario, apparently in contradiction with the observation of the cosmic microwave background (CMB) [78]. The authors suggested a nonzero Λ.

- Counts of the number of galaxies. Distant stellar objects are old enough to tell how great Λ must have been in order to affect the scale factor at earlier times. Fukugita and Turner obtained the best fit by choosing $\Omega_m + \Omega_\Lambda \sim 1$ with $\Omega_m \sim 0.1$–0.3 [79], though they failed to consider the effect of merging and chemical evolution, according to critics.

- The statistics of gravitational lensing by distant objects is also affected by space-time geometry, including the effect of Λ. The upper bound $\Omega_\Lambda \lesssim 0.8$ was obtained, though sufficient lensing data were still not yet available [80].

A more stringent result for cosmological parameters was obtained from the BOOMERANG experiment (a long-duration balloon) [81]. The CMB probes the recombination era ($z \sim 10^3$) directly. From the temperature anisotropy observed in the CMB, the density perturbation for this era was measured. Several peaks due to the acoustic property of adiabatic perturbations were observed in the spectrum of the CMB. The first peak ($\ell \sim 200$, i.e. angular scale $\approx 1°$) depends mainly on the spatial curvature Ω_k. From the BOOMERANG observation it was concluded that $\Omega_k = 0$; we have the geometry of a spatially flat universe. This has been confirmed by another experiment (MAXIMA I) [82]. If we believe that the energy density of dark matter Ω_m is ~ 0.2–0.3, as has been obtained independently by other observations, there must be another unknown and unclumpy component of energy, such as a cosmological constant, or a spatially uniform scalar field. We call this *dark energy*.

In passing we add that the inflationary scenario to be applied to the very early universe predicted a spatially flat universe, which implies that

$$\sum_i \Omega_i = 1, \qquad (5.4)$$

for the total sum of various forms of the density parameters. This "conjecture" has helped us to search for something that is supposed to fill up the left-hand side of (5.4), and eventually led to the discovery of the dark energy.

A more direct item of evidence for dark energy came finally from the observation of distant type-Ia supernovae [14, 15]. These objects serve as a standard candle allowing one to measure the cosmic distance. A supernova of this type occurs in a closed binary system with a carbon–oxygen white dwarf. A white dwarf accretes matter from a companion star and its mass approaches the Chandrasekhar mass. Then deflagration of carbon occurs, resulting in total disruption of the white dwarf. Although the physics of such a thermonuclear runaway process is complex and not well understood, we know an empirical law on the lightcurve. Applying this law to observed lightcurves and redshifts gives a relation between the apparent magnitude m and the redshift z, and thus constrains the cosmological parameters.

Both of the two observation teams, "Supernova Cosmology Project" and "High-Z Supernova Search Team," drew the same conclusion. According to Fig. 5.1, there is the relation $0.8\Omega_m - 0.6\Omega_\Lambda = -0.2 \pm 0.1$, indicating the existence of a cosmological constant. On combining this with the BOOMERANG result ($\Omega_m + \Omega_\Lambda \approx 1$), we find that $\Omega_m \sim 0.3$ and $\Omega_\Lambda \sim 0.7$.

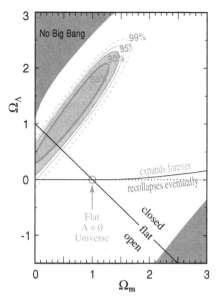

Fig. 5.1. The likely range in the $\Omega_m-\Omega_\Lambda$ plane is shown as an inclined band described by $0.8\Omega_m - 0.6\Omega_\Lambda = -0.2 \pm 0.1$, obtained from the analysis of type-Ia supernovae taken from [15]. © The American Astronomical Society (1999).

A large value of Ω_Λ obviously predicts that the universe is accelerating today, rather than decelerating as had long been believed. Needless to say, the assumed *constant* Λ could be replaced to some extent by other types of dark energy, including the effect of a dynamical field as long as it varies slowly. One of the crucial properties required is a *negative pressure*, if it is interpreted as being a fluid. Suppose that it has energy density ρ and pressure P. It has been obtained that the ratio $w \equiv P/\rho$ must be larger than a certain critical value, $w \gtrsim -0.7$, depending on the model of the dark energy [15].

Up to this time, high h_0 and low h_0 have come closer to barely "common" values of 0.6–0.7, which are not inconsistent with each other. Also the calculated age of globular clusters has decreased to $\sim 1.2 \times 10^{10}$ years. It appears as if the age problem has disappeared. That is marginally true only if the universe is open, but the problem is still there, namely we need Λ, if the universe is spatially flat, which hypothesis is supported by the anisotropy of the CMB.

In concluding this section we mention that one could imagine other explanations for the same observation avoiding the need to call for Λ, or dark energy. The observed result simply tells us that the rate of expansion in the high-z region (that is, in the past) is slower than that in our neighborhood (now); this is another expression of the accelerating universe. This is, however, true only for the model of a homogeneous universe [84].

Suppose that we live in a void of size 100 Mpc. Obviously the rate of expansion in the high-z (distant) region is slower than that in the void region, thus providing an explanation without appealing to Λ or anything like dark energy, although this requires that we live near the center of the void, which might be considered unnatural.

5.2 Quintessence

Once we accept the existence of a nonzero small Λ, we meet the second face of the problem of the cosmological constant, which was mentioned in the preceding chapter. The second face comes naturally only after the first face. The first face itself, which is related to the question of why Λ is so small compared with what the idea of unification forces upon us in a reasonable way, might have been a rather academic issue, or a conceptual problem. However, the observational "discovery" of the accelerating universe brought the first face back to being a more realistic and urgent issue. The second phase was to unfold as the "crises," as mentioned in 1989 by Weinberg [16], started mounting.

In the context of the scalar–tensor theory we had reached a successful implementation of the scenario of a decaying cosmological constant. Some of the theorists preferred, however, to appeal directly to a scalar field, without explicit reference to the scalar–tensor theory. They did not bother to introduce a nonminimal coupling term, assuming simply the standard Einstein–Hilbert term.

From the point of view of the scalar–tensor theory, these approaches, some theoretical and others phenomenological, are interpreted as working exclusively in the E frame. The exponential potential considered before is then not a unique choice. Another choice of an inverse-power potential, which was developed previously to account for the large-scale structure, has been a renewed focus of the attempts at obtaining a decaying Λ [84]. The scalar field discussed in this context is called "quintessence," though it is used in a wider sense to imply a fifth element of the universe, representing specifically an unclumped distribution with negative pressure [17]. Keeping an obvious connection with the scalar field discussed before, we denote it as σ, though other symbols such as Q and ϕ are used frequently.

We assume a potential $V(\sigma)$, in terms of which the field equation of σ is given by

$$\ddot{\sigma} + 3H\dot{\sigma} + \frac{dV}{d\sigma} = 0, \tag{5.5}$$

where we assume that σ depends only on cosmic time t, though spatial fluctuation might be reconsidered if we include further perturbation of the density with the characteristic "sound velocity" [85].

We suppress the symbol $*$, for the moment. The energy density of the scalar field is

$$\rho_\sigma = \tfrac{1}{2}\dot{\sigma}^2 + V(\sigma), \tag{5.6}$$

while the matter fluids obey the equations

$$\dot{\rho}_\mathrm{r} + 4H\rho_\mathrm{r} = 0, \tag{5.7}$$

$$\dot{\rho}_\mathrm{d} + 3H\rho_\mathrm{d} = 0, \tag{5.8}$$

for the radiation- and the dust-dominated universe, respectively.

In order to find a successful quintessence model, we need a particular type of solution of σ. Most important, ρ_σ is supposed to fall off somewhat more slowly than the ordinary energy density, which decreases like t^{-2}, as long as ρ_σ can be much smaller than ρ_r or ρ_d, and also the universe expands following a power-law of time t. If these conditions are met, ρ_σ, which starts with a much smaller initial value than that of the ordinary matter density, will catch up with the latter, thus mimicking a constant Λ to some extent, and hence causing an extra acceleration. This behavior is often called "tracking." Also the still decreasing behavior of ρ_σ can be thought of as an implementation of the idea of a decaying cosmological constant.

This tracking behavior must have lasted for a long time in order for it not to affect the nucleosynthesis and structure formation during the radiation-dominant and dust-dominant eras, respectively, subsequently becoming dominant just before the present time. This will assure that the accelerating universe at a late time is generic, thus avoiding one of the fine-tuning problems, which is sometimes called a "coincidence problem."

B. Ratra and P. J. E. Peebles examined two types of potentials [84]; exponential and inverse-power potentials. Since the exponential potential has been discussed already in subsection 4.2 of Chapter 4, we focus now on the inverse-power potential given by

$$V(\sigma) = \mu^{q+4}\sigma^{-q}, \tag{5.9}$$

where μ is again a typical mass scale of the potential. Although this potential is not renormalizable, it may appear as an effective potential for a kind of fermion condensation in a supersymmetric QCD model [86]. The $SU(N_c)$ gauge symmetry is broken by a pair condensation of N_f flavor quarks. The effective potential for a fermion-condensate field σ is given by (5.9) with $q = 2(N_c + N_\mathrm{f})/(N_c - N_\mathrm{f})$.

In the radiation- or dust-dominant era, we find an analytic solution for the scalar field:

$$\sigma = \left(\frac{q(q+2)^2}{G(q)}\right)^{1/(q+2)} \mu^{(q+4)/(q+2)} t^{2/(q+2)}, \tag{5.10}$$

and hence

$$\rho_\sigma = \left[q(q+2)^2\right]^{-q/(q+2)} G(q)^{-2/(q+2)} [2q + G(q)] \mu^{2(q+4)/(q+2)} t^{-2q/(q+2)}, \tag{5.11}$$

where $G(q)$ is a dimensionless constant defined by

$$G(q) = \begin{cases} q+6, & \text{for the radiation-dominated era,} \\ 2(q+4), & \text{for the dust-dominated era.} \end{cases} \tag{5.12}$$

From (5.11) it follows that the desired behavior of ρ_σ is obtained if

$$q > 0, \tag{5.13}$$

implying that the necessary potential is indeed an inverse-power of the quintessence field σ. We can easily show that the solution (5.10) is an attractor. Using this solution, we find the evolution of the density parameter of the scalar field to be

$$\Omega_\sigma \sim \begin{cases} a^{8/(q+2)}, & \text{for the radiation-dominated era,} \\ a^{6/(q+2)}, & \text{for the dust-dominated era.} \end{cases} \tag{5.14}$$

Any initial state satisfying the background-field-dominance will evolve into this attractor solution. If the initial value of Ω_σ is much smaller than unity, it increases with time toward the attractor solution.

It also follows from (5.11) that

$$\rho_\sigma \sim \begin{cases} a^{-4q/(q+2)}, & \text{for the radiation-dominated era,} \\ a^{-3q/(q+2)}, & \text{for the dust-dominated era.} \end{cases} \tag{5.15}$$

As is often stated, ρ_σ falls off like a negative power of the scale factor, with the exponent larger than -4 or -3 for the radiation- or dust-dominated universe, respectively.

Now we evaluate the density parameter of the scalar field more accurately. Since the two attractor solutions in each stage are independent of each other, we expect a discrepancy in the energy density of the scalar field at the time when the universe switches from radiation-dominance to dust-dominance. The difference in ρ_σ at each stage, however, appears only in the factor of $G(q)$. We can easily check that the ratio of the factor for radiation-dominance to that for dust-dominance is about 0.8 for any value of q. It then follows that there is only a slight discrepancy between the energy densities of two attractor solutions at the time of transition from radiatiation- to dust-dominance. In this way we estimate the density parameter on the basis of the analytic solution, although we need a numerical calculation if we wish to know how this attractor is reached.

The attractor solution in the dust-dominant era gives

$$\Omega_\sigma = \frac{3}{4}\left(\frac{q(q+2)^2}{2(q+4)}\right)^{-q/(q+2)} \mu^{2(q+4)/(q+2)} t_0^{4/(q+2)}, \tag{5.16}$$

where t_0 is the present time.

Fitting the observed value of the dark energy at the present epoch requires us to fix the values of μ to $\mu_{\rm obs}$ given by

$$\mu_{\rm obs} \sim \begin{cases} 6.25 \times 10^{-18}, & q = 3, \\ 8.75 \times 10^{-16}, & q = 4, \\ 4.06 \times 10^{-14}, & q = 5. \end{cases} \tag{5.17}$$

It seems that μ has to be fine-tuned. The mass scale is found to be close, however, to the electroweak energy scale. This observation suggests that the value of μ can be related to some sort of phase transition like a fermion condensation. If the value obtained in this way is different from $\mu_{\rm obs}$ shown above, we would have the accelerating universe earlier or later than the present.

The size of the scale of μ suggests that the origin of this decaying cosmological constant lies somewhere between the two kinds, primordial and vacuum energy, as discussed in section 4.1 of Chapter 4. The fermion condensation, as the suspected origin of μ, arises obviously from quantum effects, but the end result depends only on a classical object, suggesting a similarity with the primordial origin.

From a physical point of view, the inverse-power potential must have occurred at a time somewhat later than the Planck time, though the solution (5.10) itself can be extrapolated back to any earlier time from a mathematical point of view. In other words, the inverse-power potential fails to confront the first face of the problem of the cosmological constant in the conventional sense, namely the issue of how Λ as large as $\sim M_{\rm P}^4 \sim 1$ can be relaxed. One might take this as an indication for a smaller mass at unification, as suggested by the brane cosmology to be discussed in the next section.

In Fig. 5.2, we show the results of numerical calculations. This figure shows that any initial conditions satisfying radiation-dominance will

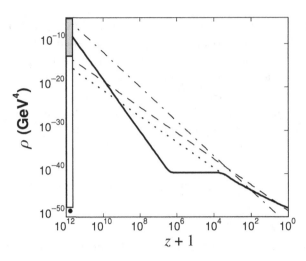

Fig. 5.2. The temporal evolution of the energy density of a tracker field with $q = 6$ (the solid line), taken from [87]. The evolutionary track of the attractor solution is shown by the dotted line. The dot–dashed and dashed lines are those of radiation and dust fluids, respectively. z is the redshift of the universe. © The American Physical Society (1999).

evolve into the attractor solution, giving an accelerating universe at the present time if we use the above value of μ.

In Fig. 5.3, we show the temporal evolution of the equation of state of the scalar field σ. From this figure, we find that, preceding the attractor solution, there is a kinetic-term-dominant stage followed by a "hesitation phase," as we call it later, featuring a plateau behavior, which lasts for a long time. This is the reason why we reached a successful quintessence model. In the kinetic-term-dominant stage, the equation of state of σ is just like that for stiff matter ($P = \rho$), hence giving ρ_σ falling off as a^{-6}, which is faster than the falling off of radiation (a^{-4}). Then the density parameter of the scalar field decreases until the attractor solution is reached.

It might appear attractive if we could start with an equipartition condition, i.e. $\rho_\sigma \sim \rho_{\rm r}/g$, where g is a degree of freedom of particles. However, as we see from Fig. 5.3, the greatest part of the energy of σ is stored in the kinetic energy; the potential energy, though there is nearly the same amount of it as there is of kinetic energy initially, would be converted into kinetic energy almost instantaneously, because the potential falls off sufficiently fast. The equipartition need not be realized naturally; it could, for example, be brought about by a fermion condensation. In this sense, we may conclude that, although any initial conditions lead to the present-day accelerating universe, the initial energy of the scalar field

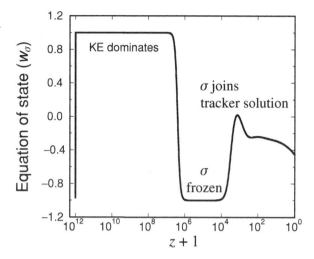

Fig. 5.3. A plot of $w_\sigma = P_\sigma/\rho_\sigma$ versus the redshift z for the solution shown in Fig. 5.2, also taken from [87]. w_σ rushes immediately toward +1 and σ becomes kinetic-energy (KE)-dominated. The field freezes and w_σ rushes toward -1. Finally, when σ rejoins the tracker solution, w_σ increases, briefly oscillates and settles into the tracker value. © The American Physical Society (1999).

is likely much smaller than the radiation energy if we start with the natural equipartition condition. Many modified models have been proposed without confirming their origins.

It nevertheless seems fair to conclude that the idea of quintessence provides a reasonably successful scenario for a decaying cosmological constant. The attractor nature of the solution naturally explains why a scalar-field-dominance is realized for late times.

5.3 Quintessence in the brane world

It still seems that fine-tuning the parameter μ to some extent is unavoidable in conventional quintessence models. We are going to show that the situation can be improved by assuming that we live in the brane universe, which provides a much wider and more natural range for initial values. The five-dimensional Planck mass and the mass scale of the potential will be determined roughly to be of the order of 10^{5-6} GeV, remarkably different from the conventional view of the hierarchy problem.

Although the initial energy of the scalar field could be the same as that of the radiation fluid, the largest contribution comes from its kinetic energy. If the kinetic term dominates, then the scalar field behaves like a massless field, which is equivalent to stiff matter. Its energy density then drops as a^{-6}, much faster than does the radiation energy. The kinetic energy keeps falling off in the radiation-dominant universe until a tracking solution is finally reached.

Furthermore, from the condition that the mass scale of the potential must be tuned for a successful quintessence scenario, the potential term cannot be so large. It then follows that an equipartition condition, which may be expected in the early stage of the universe, seems unlikely.

Another unsatisfactory point is that, if the universe starts in a scalar-field-dominant state, the radiation-dominant universe is never recovered, making it unlikely that we would find the present-day universe as we see it. We may expect to find a way out by exploring a quintessence model in a brane world.

5.3.1 A scalar field in the brane world

The 00-component of Einstein's equation in Robertson–Walker space-time in our brane world, with $k = 0$ and $\Lambda = 0$, for simplicity, is given by [88,89]

$$3H^2 = \rho + \frac{1}{12(M_\mathrm{P}^{(5)})^6}\rho^2 + \frac{\mathcal{K}}{a^4}, \qquad (5.18)$$

where ρ is the energy density of matter fields. $\mathcal{K}$ is a constant, which describes "dark" radiation coming from $E_{\mu\nu}$ (some components of the five-dimensional Weyl tensor) [89]. We notice the occurrence of the two new terms, i.e. a quadratic term of energy density and "dark" radiation, which are unique to the brane scenario. See Appendix L for more details. We concentrate on the dynamical behavior of the scalar field, though effects on the early history of the universe have been discussed widely.

As for matter fields on the brane, we assume a scalar field σ as well as the conventional radiation and matter fluids, i.e. $\rho = \rho_\sigma + \rho_\mathrm{r} + \rho_\mathrm{m}$, where $\rho_\sigma, \rho_\mathrm{r}$, and ρ_m are the energy densities of the scalar field σ, of radiation and of dust matter, respectively. (Only throughout this section, we follow a custom in the discipline within which context our original papers were prepared, namely that of using the subscript m instead of d, for dust matter.) Although we can consider a five-dimensional scalar field living in the bulk [34,89], we shall focus only on a four-dimensional scalar field confined on the brane. The origin of this scalar field might be a condensation of some fermions confined on the brane, as discussed in the preceding section. Since the energy of each field on the brane is conserved in the present model, we find the same dynamical equation for a four-dimensional scalar field as for a conventional one (5.5) without considering the brane.

As the universe expands, the energy density decreases. This means that the quadratic term was important only in the early stage of the universe. On comparing the two terms (the conventional energy density term (ρ) and the quadratic one (ρ^2)), we find that the quadratic term dominates when

$$\rho > \rho_c \equiv 12(M_{\mathrm{P}}^{(5)})^6. \tag{5.19}$$

When the quadratic term is dominant, the law regulating the expansion of the universe is modified. For example, in the radiation-dominant era, i.e. $\rho_{\mathrm{r}} \gg \rho_{\mathrm{m}}, \rho_\sigma$, the universe expands as $a \sim t^{1/4}$, in contrast to $t^{1/2}$ in the conventional radiation-dominant era.

Since we are interested in a quintessence scenario and the ensuing dynamical behavior of the scalar field, we shall calculate the density parameter of the scalar field (Ω_σ). Ignoring dark radiation ($\mathcal{K} = 0$), we find

$$\dot{\Omega}_\sigma = \frac{H\rho_{\mathrm{r}}}{\rho^2}\left(4V(\sigma) - \dot{\sigma}^2\right). \tag{5.20}$$

When the quadratic term is dominant in the early stage, the Hubble expansion rate decreases. As a result, the friction term in (5.5) becomes small and a kinetic term will play a more important role in the energy density of σ. Consequently, if $\dot{\sigma}^2 > 4V(\sigma)$, the density parameter of the scalar field will decrease with time. This feature may be particularly important in a quintessence scenario, because most quintessence models assume very small potential energy in its early stage for a successful scenario. We will show that this interesting feature is found in some quintessence models and that the initial conditions for a successful scenario become much wider in a brane world.

In what follows, assuming an inverse-power-law potential, which is most interesting in the present quintessence scenario, we investigate the dynamical behavior of the scalar field. Since the behavior has been studied well in the conventional universe [17, 87, 91, 92], we turn our attention mainly to the period in which the quadratic term (ρ^2) is dominant.

In the radiation-dominant universe, the scale factor expands as $a \sim t^{1/4}$ and the equation for the scalar field (5.5) is now

$$\ddot{\sigma} + \frac{3}{4t}\dot{\sigma} - q\mu^{q+4}\sigma^{-q-1} = 0. \tag{5.21}$$

We find an analytic solution for $q < 6$ [93]:

$$\sigma = \sigma_0\left(\frac{t}{t_0}\right)^{2/(q+2)}, \tag{5.22}$$

with

$$\sigma_0^{q+2} = \frac{2q(q+2)^2}{6-q}\mu^{q+4}t_0^2, \tag{5.23}$$

where t_0 and σ_0 are integration constants. The energy density of the scalar field evolves as

$$\rho_\sigma = \frac{3q(q+2)}{6-q}V_0\left(\frac{t}{t_0}\right)^{-2q/(q+2)} = \frac{3q(q+2)}{6-q}V_0\left(\frac{a}{a_0}\right)^{-8q/(q+2)}, \tag{5.24}$$

where $V_0 = V(\sigma_0)$ and $a_0 = a(t_0)$. For the density parameter Ω_σ, we find

$$\Omega_\sigma = \frac{\rho_\sigma}{\rho_r + \rho_\sigma} \approx \frac{\rho_\sigma}{\rho_r} = \Omega_\sigma^{(0)}\left(\frac{a}{a_0}\right)^{4(2-q)/(q+2)}, \tag{5.25}$$

where

$$\Omega_\sigma^{(0)} = \frac{3q(q+2)}{6-q}\frac{V_0}{\rho_r(t_0)}, \tag{5.26}$$

because $\rho_r \propto a^{-4}$.

If $q > 2$, just contrary to the tracking solution, the scalar-field energy decreases faster than does that of the radiation. This result is confirmed by (5.20) with the fact that the kinetic energy of a scalar field $\rho_\sigma^{(K)} = \dot{\sigma}^2/2$ turns out to be larger than $4V$ because $\rho_\sigma^{(K)}/(4V) = 2q/(6-q)$.

If $q = 2$, the scalar-field energy drops at the same rate as does that of the radiation until the conventional universe is recovered. This is the so-called scaling solution. If $q < 2$, ρ_σ, the scalar-field energy, decreases more slowly than does the radiation energy, and eventually the scalar field dominates over the radiation.

If $q \geq 6$, there is no asymptotic solution for which the kinetic term balances with the potential term, resulting in a kinetic-term-dominant solution asymptotically.

In the conventional universe, once a quintessence field dominates, radiation-dominance will not be obtained. In the brane cosmology, however, we can easily show that a radiation-dominant era is recovered if $q \geq 2$. We summarize the obtained analytic solutions and their fates in Table 5.1.

5.3.2 Quintessence in the brane world

Now we are ready to discuss a quintessence scenario in the brane world. We assume that $2 < q < 6$. Using two attractor solutions (one in the ρ^2-dominant stage and the other in the conventional universe), we present a successful and natural scenario [93]. Since the quintessence solution in the conventional-universe model is an attractor, our solution should also

Table 5.1. The fate of the scalar field for each initial condition, where S and R denote scalar-field-dominance ($\rho_\sigma \gg \rho_r$) and radiation-dominance ($\rho_r \gg \rho_\sigma$), respectively. The "scaling" means the scaling solution ($\Omega_\sigma = \text{constant}$). The asymptotic behavior $\rho_\sigma^{(P)}/\rho_\sigma^{(K)} = \text{constant}$ is described by "Constant," while the kinetic dominance $\rho_\sigma^{(K)} \gg \rho_\sigma^{(P)}$ is denoted by "Kinetic." "PL" is power-law. "DE" is decelerating power-law expansion [93].

q	μ	Initial	Fate	Feature
$q < 2$	Any value	S	S	Inflation
		R		
$q = 2$	$\mu/m_5 > 40^{1/6}$		S	PL inflation
	$4^{1/6} < \mu/m_5 < 40^{1/6}$	S		DE
	$\mu/m_5 < 4^{1/6}$		R	Scaling
	Any value	R		
$2 < q < 6$	Any value	S	R	Constant
		R		
$q > 6$	Any value	S	R	Kinetic
		R		

recover the same trajectory after the quadratic term has decreased to a very small value. The main difference is that we can include not only radiation-dominant initial conditions but also scalar-field-dominant initial conditions for a successful scenario.

If we ignore the above small discrepancies at t_c, the time corresponding to the transition (5.19), and t_{eq}, we can estimate the density parameter Ω_σ from (5.25) and (5.14):

$$\Omega_\sigma = \Omega_\sigma^{(s)} \times \left(\frac{a_c}{a_s}\right)^{-(q-2)/(2q+2)} \times \left(\frac{a_{eq}}{a_c}\right)^{8/(q+2)} \times \left(\frac{a_0}{a_{eq}}\right)^{6/(q+2)}$$

$$= \Omega_\sigma^{(s)} \times \left(\frac{T_s}{T_c}\right)^{-(q-2)/(2q+2)} \times \left(\frac{T_c}{T_{eq}}\right)^{8/(q+2)} \times \left(\frac{T_{eq}}{T_0}\right)^{6/(q+2)}, \quad (5.27)$$

where $\Omega_\sigma^{(s)}$ is the density parameter when the attractor solution is reached. For a successful quintessence scenario, we require that the present-day value of the density parameter of the scalar field be $\Omega_\sigma \sim 0.7$.

In what follows, we discuss constraints for a natural and successful quintessence. We first consider three constraints: nucleosynthesis, matter-dominance at the decoupling time, and natural initial conditions, in this order.

(a) Nucleosynthesis One of the most successful results of the big-bang standard cosmology is a natural explanation of the present-day amounts of light elements. In any cosmological model, a successful model of nucleosynthesis provides a necessary constraint, which may be most stringent. During nucleosynthesis, the universe must have expanded as the conventional radiation-dominant universe. Therefore, the transition from the ρ^2-dominant stage to the conventional universe must take place before nucleosynthesis. This constraint gives a lower bound for the value of m_5. Introducing two temperatures, T_c and T_{NS}, which correspond to those at the transition time t_c and at the nucleosynthesis time t_{NS}, respectively, we impose a constraint $T_c > T_{NS}$, which implies that

$$\rho_c > \rho_r(t_c) = \frac{\pi^2}{30} g_{NS} T_{NS}^4, \tag{5.28}$$

where g_{NS} is the degree of freedom of particles at nucleosynthesis. Since $\rho_c = 12(M_P^{(5)})^6$ and $T_{NS} \sim 1\,\text{MeV}$, the constraint (5.28) yields

$$M_P^{(5)} > 1.6 \times 10^4 (g_{NS}/100)^{1/6} \times (T_{NS}/1\,\text{MeV})^{2/3}\,\text{GeV}. \tag{5.29}$$

In the reduced Planckian unit system, this constraint can be written as $M_P^{(5)} > 10^{-14}$. If the Randall–Sundrum type-II model is a fundamental theory, in order to recover the Newtonian force above the 1-mm scale in the brane world, the five-dimensional Planck mass is constrained as $M_P^{(5)} \geq 10^8\,\text{GeV} \sim 4 \times 10^{-11}$ [33], which would be a stronger constraint. However, the Randall–Sundrum type-II model could be an effective theory, derived from more fundamental higher-dimensional theories such as the Hořava–Witten theory [29]. Thus, we adopt the above constraint here.

(b) Matter-dominance at the decoupling time From the observation of the CMB, we have information on the universe at $T \sim 4000\,\text{K}$, from which we expect that the inhomogeneity of the universe was about 10^{-5}. In order to form some structure from the decoupling time to the present, the energy density of the matter fluid should be larger than that of the "dark energy" (that of the scalar field) by a few orders of magnitude at the decoupling time t_{dec}.

The energy density of the matter fluid at the decoupling time is given as

$$\rho_m(t_{dec}) = \rho_m(t_{eq}) \left(\frac{a_{eq}}{a_{dec}}\right)^3 = \rho_r(t_{eq}) \left(\frac{T_{dec}}{T_{eq}}\right)^3$$

$$= \left(\frac{\pi^2}{30}\right) g_{eq} T_{eq}^4 \left(\frac{T_{dec}}{T_{eq}}\right)^3. \tag{5.30}$$

The energy density of radiation is estimated as follows. From Einstein's equation in the radiation-dominant era $(H = \rho_r/(6(M_P^{(5)})^3))$, we find $\rho_r = 3(M_P^{(5)})^3/(2t)$ in the ρ^2-dominant stage. In the conventional stage $(\rho_r \sim t^{-2})$, because $t_c = 1/(8(M_P^{(5)})^3)$ from the definition (5.19), we find

$$\rho_r = \frac{3}{16}t^{-2}, \quad \text{for} \ \ t_c < t < t_{eq}. \tag{5.31}$$

As for the energy density of σ, assuming that the attractor (5.11) with (5.12) is reached and using (5.31), we estimate $\rho_\sigma(t_{dec})$ in terms of μ, T_{eq}, and T_{dec} for each q.

T_{eq} is about $10^4 \, \text{K} < T_{eq} < 10^5 \, \text{K}$. In order to impose the most stringent constraint, we adopt $T_{eq} = 10^4 \, \text{K}$ here. Setting $T_{dec} = 4000 \, \text{K}$ and $T_{eq} = 10^4 \, \text{K}$, from the constraint of $\rho_m(t_{dec}) > \rho_\sigma(t_{dec})$, we find the upper bounds for the value of μ:

$$\mu < \begin{cases} 1.23 \times 10^{-16}, & \text{for } q = 3, \\ 1.31 \times 10^{-14}, & \text{for } q = 4, \\ 2.40 \times 10^{-13}, & \text{for } q = 5. \end{cases} \tag{5.32}$$

Although the value of μ is fixed if the scalar field dominates now $(\Omega_\sigma(t_0) \sim 0.7)$ as (5.17), since we do not know the value of μ from the viewpoint of particle physics, we shall let its value be free.

(c) Initial conditions Since a quintessence solution is an attractor, we may not need to worry about initial conditions. In fact, the conventional quintessence will be recovered even in the present model. What may be better in the present model is that the basin of attraction is much larger. In particular, the conventional quintessence might not work if a scalar field dominates initially, but it will still work in the present model.

Nevertheless, since we assume that a quintessence field σ is confined on the brane, we have one stringent constraint, that is, all energy scales including its potential should be smaller than the five-dimensional Planck scale $M_P^{(5)}$;

$$\mu \leq M_P^{(5)}. \tag{5.33}$$

The maximal possible energy density of radiation is about $(M_P^{(5)})^4$.

With the constraints (a) and (b), we restrict two unknown parameters; $M_P^{(5)}$ and μ. In Fig. 5.4, we plot these constraints as three solid lines in the μ–$M_P^{(5)}$ parameter space for $q = 4$ and 5.

Then, if μ is fixed to give scalar-field-dominance right now, (5.17), the five-dimensional Planck mass $M_P^{(5)}$ is limited. For example, for $q = 5$, if $\mu < 10^{-14}$, the constraint for $M_P^{(5)}$ comes just from nucleosynthesis, whereas if $\mu > 10^{-14}$, $M_P^{(5)} > \mu$, which is a stronger constraint than that from nucleosynthesis.

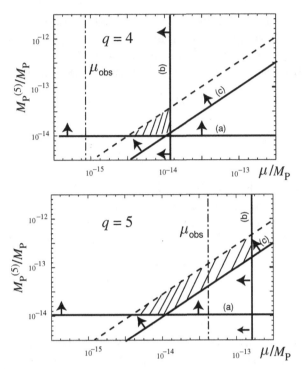

Fig. 5.4. Constraints in the μ–$M_{\mathrm{P}}^{(5)}$ parameter space. The three solid lines are for (a) nucleosynthesis, (b) matter-dominance at the decoupling time, and (c) an energy scale smaller than the five-dimensional Planck mass. The observation ($\Omega_\sigma \sim 0.7$) fixes the value of μ (μ_{obs}), which is given by the vertical dot–dashed lines. If we assume that the potential energy of the scalar field is initially not very small compared with its kinetic energy, a successful quintessence is possible for the shaded region [93]. © The American Physical Society (1999).

(d) The equipartition condition We try to invoke a further constraint that could be derived from natural initial conditions. What would those natural conditions be for the scalar field? One of the plausible conditions is an equipartition for each energy density. It will then follow that the radiation energy is larger than that of the scalar field because the degree of freedom of all particles g is larger than that of the scalar field. How about the ratio of the kinetic energy to the total energy of the scalar field? In the conventional quintessence scenario, the potential energy should be initially much smaller than the kinetic energy. In the present model, however, this is not the case, because the attractor solution in the ρ^2-dominant stage

reduces the density parameter of the scalar field. This allows us to impose natural initial conditions for the scalar field.

To be more concrete, we focus on the potential due to fermion condensation. After our three-dimensional brane world has been created, a fermion pair is condensed by a symmetry-breaking mechanism and it behaves as a scalar field with a potential $V(\sigma)$. We expect that the potential term should play an important role right from the initial stage.

We consider the choices $q = 3, 4$, and 5. The initial conditions for the scalar field are classified into the following three cases. First, if the kinetic and potential energies of the scalar field are of the same order of magnitude, the attractor solution is reached soon, and then we expect $\Omega_\sigma^{(s)} \sim 0(g_s^{-1})$, where g_s is a degree of freedom of particles at a_s, for which the attractor solution is reached. Secondly, if the potential energy is dominant, it will not change so much before reaching the attractor solution, and then we expect that $\Omega_\sigma^{(s)} \gtrsim 0(g_s^{-1})$. Thirdly, if the kinetic energy is larger than the potential energy, it will decay soon, finding an attractor solution, so $\Omega_\sigma^{(s)} \lesssim 0(g_s^{-1})$, unless the kinetic energy is too dominant, which we do not assume here. In this way, we come to the conclusion that a natural initial condition predicts $\Omega_\sigma^{(s)} \sim 0(g_s^{-1})$.

From the evolution equations of the energy densities of the scalar field and of the radiation, we find

$$\Omega_\sigma^{(s)} = F(q) \left(\frac{\mu}{M_P^{(5)}} \right)^{2(q+4)/(q+2)} (M_P^{(5)} t_s)^{-2q/(q+2)}, \qquad (5.34)$$

where

$$F(q) = 2^{(q+4)/(q+2)} q^{-q/(q+2)} (q+2)^{-(q-2)/(q+2)} (6-q)^{-2/(q+2)}/3. \quad (5.35)$$

Since $\rho_r(t_s) \lesssim (M_P^{(5)})^4$, we have a constraint that $M_P^{(5)} t_s \gtrsim 1.5$ from the equation for the radiation energy, $\rho_r = \frac{3}{2}(M_P^{(5)})^3 t^{-1}$. This constraint combined with $\Omega_\sigma^{(s)} \sim 0(g_s^{-1})$ and $g_s \sim 10^3$ gives the lower bound for $\mu/M_P^{(5)}$:

$$\mu \gtrsim \begin{cases} 0.131 M_P^{(5)}, & \text{for } q = 3, \\ 0.140 M_P^{(5)}, & \text{for } q = 4, \\ 0.146 M_P^{(5)}, & \text{for } q = 5. \end{cases} \qquad (5.36)$$

With the previous constraint $\mu < M_P^{(5)}$, we find a narrow strip in the μ–$M_P^{(5)}$ parameter space, which is shown as the shaded regions in Fig. 5.4.

The allowed region becomes smaller for smaller values of q. In particular, we find that no region is allowed for $q \leq 3$. Therefore, the present

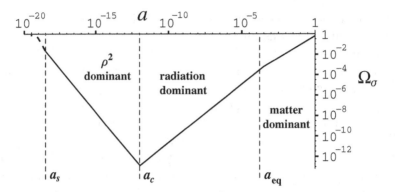

Fig. 5.5. The time-variation of Ω_σ in terms of the scale factor a, with its present-day value normalized to unity, for the model with $V = \mu^9 \sigma^{-5}$. We chose $\Omega_\sigma^{(s)} = 0.01$, $T_s = \mu \sim 2 \times 10^5\,\text{GeV}$, and $M_P^{(5)} \sim 2 \times 10^6\,\text{GeV}$ [93]. © The American Physical Society (1999).

model prefers a rather large value of q. In Fig. 5.5, we show an example of the time-variation of Ω_σ for the choice of $q = 5$ with $\Omega_\sigma^{(s)} = 0.01$ and $T_s = \mu \sim 2 \times 10^5\,\text{GeV}$, which is smaller than $M_P^{(5)} \sim 2 \times 10^6\,\text{GeV}$.

With the values (5.17), we find that there are no natural ranges for $q \leq 4$. For $q \gtrsim 5$, the five-dimensional Planck scale is strictly constrained from the observation because the allowed region is very narrow. For $q = 5$, for example, we find

$$4.06 \times 10^{-14} \lesssim M_P^{(5)} \lesssim 2.78 \times 10^{-13}. \tag{5.37}$$

Achieving a reasonably consistent way of understanding the problem of the cosmological constant in terms of the inverse-power potential might be a major success of the brane cosmology, which is still viewed as somewhat hypothetical.

5.4 Scalar–tensor theory

5.4.1 Hesitation behavior

In the E frame of the prototype BD model with Λ included, the potential of the scalar field σ is exponential, one of the examples of a *rapidly falling* potential. With this unique potential we have a generic behavior that seems to provide nearly the same effect as that of a constant Λ, as will be shown explicitly.

We may have a phase of kinetic-term-dominance, represented by

$$K = \tfrac{1}{2}\dot\sigma \gg V(\sigma). \tag{5.38}$$

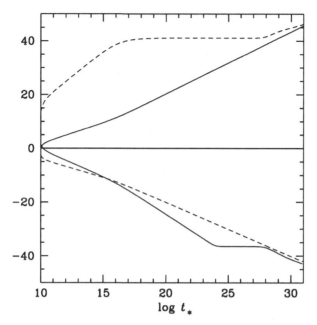

Fig. 5.6. An example of hesitation behavior. The solid curve in the upper half of the plot shows $2\ln a_*$, while the dashed curve represents 2σ. In the lower half of the plot, the dashed and solid curves are for $\log(10^{20}\rho_*)$ and $\log(10^{20}\rho_\sigma)$, respectively. We chose $\zeta = 1.5823$, the same as will be used in the two-scalar model in the next subsection. The initial values at $\log t_{*1} = 10$ are given by $\sigma_1 = 6.754\,42$ and $\dot{\sigma}_1 = 0$, while the matter density, which is assumed to be radiation-dominated, is 3.7352×10^{-23}.

We also consider that the matter density ρ_* is negligibly small;

$$\rho_\sigma \approx K \gg \rho_{*\mathrm{m}}. \tag{5.39}$$

This might have occurred if the re-heating process after the inflation had not been strong enough, for example. This is also true after a mini-inflation, as we will see. In Fig. 5.6, we chose the parameters such that we begin with this kinetic-term-dominance.

In this situation we have the following approximate equations:

$$3H_*^2 = K, \tag{5.40}$$
$$\ddot{\sigma} + 3H_*\dot{\sigma} = 0. \tag{5.41}$$

We immediately find the solution

$$H_* = \tfrac{1}{3}t_*^{-1}, \tag{5.42}$$

$$\dot{\sigma} = \sqrt{\frac{2}{3}}t_*^{-1}. \tag{5.43}$$

The only ambiguities are the origin of the time t_* and the sign of $\dot{\sigma}$, which are irrelevant. Notice that only the expanding universe with $H_* = +\frac{1}{3}t_*^{-1}$ follows, instead of a negative H_*. This is related to the fact that a spatially uniform scalar field without potential is equivalent to a fluid with the equation of state $P = \rho$, which is known to yield $H_* = \frac{1}{3}t_*^{-1}$.

By integrating (5.43) we obtain

$$\sigma(t_*) = \sqrt{\frac{2}{3}}\ln t_* + \bar{\sigma}, \tag{5.44}$$

where $\bar{\sigma}$ is a constant. We chose σ to increase with time. Equation (5.44) can also be put into the form

$$e^{\sigma - \bar{\sigma}} = t_*^{\sqrt{2/3}}, \tag{5.45}$$

which can be used to calculate

$$V(\sigma) = \Lambda e^{-4\zeta\bar{\sigma}} t_*^{-4\zeta\sqrt{2/3}}, \tag{5.46}$$

verifying that the potential term falls off much faster than K:

$$K = \frac{1}{3}t_*^{-2}. \tag{5.47}$$

Now take a small density ρ_* into account. The covariant conservation law in the radiation-dominated era is given by

$$\dot{\rho}_* + 4H_*\rho_* = 0. \tag{5.48}$$

Substituting from (5.42) gives

$$\rho_*(t_*) = \rho_{*0}t_*^{-4/3}. \tag{5.49}$$

On comparing this with (5.47) we find that ρ_* catches up with $K \approx \rho_\sigma$ at time t_{*2};

$$t_{*2} = (3\rho_{*0})^{-3/2}. \tag{5.50}$$

This catch-up time can be seen around $\log t_{*2} \approx 15$ in Fig. 5.6. After this time, the universe enters a radiation-dominated era;

$$H_* = \frac{1}{2}t_*^{-1}, \tag{5.51}$$

$$\rho_* \sim t_*^{-2}. \tag{5.52}$$

By substituting (5.51) into (5.41) we now obtain

$$\sigma = \sigma_3 - 2t_*^{-1/2}, \tag{5.53}$$

$$K \sim t_*^{-3}. \tag{5.54}$$

We find that σ tends to a finite value σ_3, which is determined by the conditions at t_{*2}, with ever decreasing speed. This will be called a *hesitation* period. One may also call it a *hibernation*. The curve of σ and that of ρ_σ level off, providing a *plateau*, playing, it is hoped, the same role as that of a cosmological constant. In Fig. 5.6, the plateau of σ begins around $\log t_{*3} \approx 18$, whereas that of ρ_σ starts somewhat later, at $\log t_{*4} \approx 24$; the kinetic term had still decreased in a way not precisely visible in the curve of σ. How this behavior is related to the observation depends, however, on how it ends.

During this hesitation phase, the σ energy stays at the value $\rho_\sigma(\sigma_4) = V(\sigma_4)$ which depends on the previous history. This is obviously very small, but $\rho_*(t_*) \sim t_*^{-2}$ will decrease, and eventually will become as small as $V(\sigma_4)$, as indicated clearly at around the time $\log t_{*5} \approx 28$ in the example of Fig. 5.6. Then the period of radiation-dominance comes to an end. No kinetic-term-dominance in ρ_σ remains true either. We must then include all the terms, K, V, and ρ_*, on equal footing, coming back to the usual development of section 4.4 in Chapter 4. In other words, a transient solution for the hesitation period will enter the attractor solution in the asymptotic phase, which we discussed before.

We are interested in how transition from the hesitation behavior to the attractor behavior takes place. If the plateau in the ρ_σ curve crosses ρ_*, it may represent a situation $\Omega_\Lambda \gtrsim 0.5$, which looks like the result of recent observations. Detailed numerical simulations seem to be needed in order to see how the transition proceeds. According to the results of past studies, this does not appear to be the case.

Precisely the same kind of behavior was encountered in Fig. 5.2, and described by the term "freezing," for the quintessence with an inverse-power potential and an exponential of the inverse field potential. The same behavior was due to an "overdamped" scalar field according to Carroll [94]. These potentials are also examples of rapidly falling potentials.

According to some numerical calculations on the above-mentioned potentials, the σ energy was shown to surpass the matter density, thus implementing the condition $\Omega_\Lambda \gtrsim 0.5$. The same favorable result no longer occurs with the exponential potential, as was shown by Ferreira and Joyce [95].

As in a typical example of Fig. 5.6, σ tends to move ahead as the hesitation phase comes close to the end. The force exerted on σ from the exponential potential becomes relatively significant as the value of ρ_σ becomes comparable to $\rho_* \sim t_*^{-2}$, forcing the plateau of ρ_σ to bend downward. We have never seen any examples in which ρ_σ extends to thrust into ρ_*, as a truly constant Λ would have done. It is as if σ were seduced to enter the asymptotic behavior as quickly as possible, or *prematurely*.

From our experience, any realistic hope of reproducing results with Ω_Λ considerably larger than 0.5 disappears in this way, to our disappointment. It nevertheless maintains the unique tendency that a Λ-like term has Ω_Λ basically of the order of unity whenever it occurs at all, which provides an automatic answer to the coincidence problem, at least partially.

It seems that the reason for this "failure" is our choice of $\zeta > \frac{1}{2}$, (4.112), or $\zeta > \sqrt{3}/4$, (4.113), for positivity of the matter density. We may rather choose a vacuum-dominated solution for which the asymptotic solution is given by (4.115) with (4.116). This might in fact be a better approximation for an accelerating universe if

$$\zeta^2 < \tfrac{1}{8}, \tag{5.55}$$

which implies then $p > 1$, a power-law inflation.

To be more realistic, we may expect a "switching" during around the present epoch from $\rho_* \gg \rho_\sigma$ to $\rho_\sigma \gg \rho_*$. This implies, however, that the initial value of ρ_σ must have been prepared perhaps many orders of magnitude smaller than t_*^{-2}, requiring a fine-tuning.

As a possible way out, the required value might have been prepared as the hesitation behavior. We recall that this has to be a consequence of the preceding kinetic-term-dominance over the potential energy, as mentioned before. From (5.46) we find that the condition to be satisfied is given by

$$4\zeta\sqrt{\frac{2}{3}} > 2, \tag{5.56}$$

which translates into

$$\zeta^2 > \tfrac{3}{8}, \tag{5.57}$$

which is unfortunately in contradiction with (5.55).

5.4.2 A two-scalar model

We want to inherit a favorable reply to the coincidence problem as above, yet arrive at a more satisfactory outcome regarding Ω_Λ. In order to keep σ from being lured away toward the end of the hesitation period, we probably must confine σ by setting up a barrier, which, however, has to be of finite height, thus allowing σ to escape relatively soon. Otherwise, the bottom of the potential barrier will serve as another cosmological constant.

After having searched for a candidate model on a trial-and-error basis, we decided to introduce another scalar field, conveniently called $\chi(x)$, and came across a potential in the E frame [96–98]:

$$V(\sigma, \chi) = e^{-4\zeta\sigma}\left\{\Lambda + \tfrac{1}{2}m^2\chi^2\left[1 + \gamma\sin(\kappa\sigma)\right]\right\}, \tag{5.58}$$

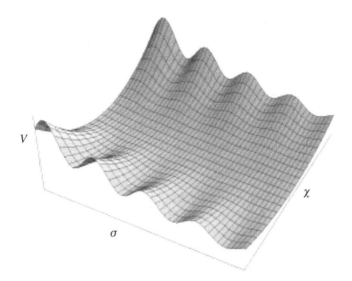

Fig. 5.7. The potential $V(\sigma, \chi)$ given by (5.58). Along the central valley with $\chi = 0$, the potential reduces to the simpler behavior $\Lambda e^{-4\zeta\sigma}$ as given by (4.92), but with $\chi \neq 0$ it exhibits an oscillation in the σ direction.

with m, γ, and κ constants chosen to be not very much different from unity in the reduced Planckian unit system. They are shown graphically in Fig. 5.7. Notice that, with $\chi = 0$, the potential (5.58) goes back to the potential in the prototype BD model with Λ as given by (4.92).

The field equations are also derived:

$$3b'^{\,2} = t_*^2(\rho_s + \rho_*), \tag{5.59}$$

$$\sigma'' + (3b' - 1)\sigma' + t_*^2 e^{-4\zeta\sigma}\left(-4\zeta\tilde{V} + \frac{\partial\tilde{V}}{\partial\sigma}\right) = \eta_{\mathrm{dm}}\zeta_{\mathrm{dm}}t_*^2\rho_*, \tag{5.60}$$

$$\chi'' + (3b' - 1)\chi' + t_*^2 e^{-4\zeta\sigma}\frac{\partial\tilde{V}}{\partial\chi} = 0, \tag{5.61}$$

$$\rho_*' + (4 - \eta_d)b'\rho_* = -\eta_{\mathrm{dm}}\zeta_{\mathrm{dm}}t_*\sigma'\rho_*, \tag{5.62}$$

where $\tilde{V}$ is the portion that multiplies $e^{-4\zeta\sigma}$ on the right-hand side of (5.58), also with

$$\rho_s = \tfrac{1}{2}\dot{\sigma}^2 + \tfrac{1}{2}\dot{\chi}^2 + V(\sigma, \chi), \tag{5.63}$$

for the energy density of the scalar fields, where a prime implies a derivative with respect to the new time variable τ, defined by

$$\tau = \ln t_*. \tag{5.64}$$

We also use

$$b = \ln a_*, \tag{5.65}$$

from which it follows that

$$b' = \frac{db}{d\tau} = t_* H_*. \tag{5.66}$$

Using τ in place of t_* has an added advantage in that the coefficient of the cosmological frictional term is a constant if the original scale factor a_* expands according to a power law; hence $H_* \sim t_*^{-1}$. It is even more interesting to find that the solutions turn out to exhibit important structures that are almost equally spaced if they are plotted against τ, as we will soon see.

We seek solutions with spatially uniform σ and χ. In (5.60) and (5.62), we use $\eta_{\rm dm} = 0$ and 1 for the radiation- and dust-dominated universe, respectively, with $\zeta_{\rm dm}$ the averaged value of ζ for the dust matter. The occurrence of this "effective" ζ is a consequence of departure from the prototype BD model as discussed in subsection 4.3.3.

An anomaly-type QCD calculation, which will be outlined in section 6.4 of Chapter 6, yields $\zeta_{\rm q}/\zeta = (5/\pi)\alpha_{\rm QCD} \approx 0.3$ for the quark with $\alpha_{\rm QCD} \approx 0.2$, the QCD counterpart of the fine-structure constant. We then find $\zeta_{\rm N}/\zeta \approx \langle \sum m_{\rm q}\bar{q}_* q_* \rangle/m_{\rm N} \approx 0.02$ for the nucleon. For the dust matter we could choose $\zeta_{\rm dm} = \zeta_{\rm N}$, but more likely $\zeta_{\rm dm}$ would be smaller, because the baryonic component is only a minor part of the whole of the dust matter. For this reason we ignore $\zeta_{\rm dm}$ in this subsection, confining ourselves to classical effects.

The structure of (5.58) might appear somewhat complicated and artificial, but is not entirely unnatural in view of the fact that a sinusoidal potential has been discussed in the literature [99, 100], though only in the context of a single scalar field. One might still ask whether this potential can be derived from any theoretical models at a more fundamental level. We postpone the answer, for now, because it still seems too early to expect theoretical models of unification to entail unambiguous predictions about the realistic world. We expect instead that any success in the phenomenological approach to such a nontrivial issue as the problem of cosmological constant would provide an important clue for theoretical efforts toward unification. From this point of view, we will concentrate on showing how the potential provides us with a reasonable way of understanding the results of recent observations. On taking this attitude, we first offer an example of such a solution.

In the top display of Fig. 5.8, we notice that the scale factor (its logarithm) features two rises against a smooth background behavior. Each of the rises implies a mini-inflation that lasts only for a relatively "short" period. We reasonably fine-tuned parameters in such a way that one of

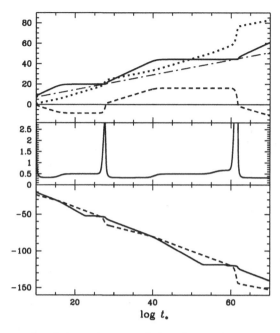

Fig. 5.8. An example of the solution of the classical equation for $\zeta_{\mathrm{dm}} = 0$. Upper diagram: $b = \ln a_*$ (dotted), σ (solid), and 2χ (dashed) are plotted against $\log t_{*1}$. The present epoch corresponds to $\log t_* = 60.1$–60.2; the primordial nucleosynthesis must have taken place at $\log t_* \sim 45$. The parameters are $\Lambda = 1, \zeta = 1.5823, m = 4.75, \gamma = 0.8$, and $\kappa = 10$. The initial values at $t_1 = 10^{10}$ are $\sigma_1 = 6.7544, \sigma_1' = 0$ (a prime implies differentiation with respect to $\tau = \ln t_*$), $\chi_1 = 0.21, \chi_1' = -0.005, \rho_{1\mathrm{rad}} = 3.7352 \times 10^{-23}$, and $\rho_{1\mathrm{dust}} = 4.0 \times 10^{-45}$. The dashed–dotted straight line represents the asymptote of σ given by $\tau/(2\zeta)$. Notice the long plateaus of σ and χ, and their rapid changes during relatively "short" periods. Middle diagram: $p_* = b' = t_* H_*$ for an effective exponent in the local power-law expansion $a_* \sim t_*^{p_*}$ of the universe. Notable leveling-offs can be seen at 0.333, 0.5, and 0.667, corresponding to the epochs dominated by the kinetic terms of the scalar fields, the radiation matter, and the dust matter, respectively. Lower diagram: $\log \rho_s$ (solid), the total energy density of the σ–χ system, and $\log \rho_*$ (dashed), the matter energy density. Notice the "interlacing" pattern of ρ_s and ρ_*, still obeying $\sim t_*^{-2}$ as an overall behavior. Nearly flat plateaus of ρ_s occur before ρ_s overtakes ρ_*, hence with Ω_Λ passing through 0.5. One of the main effects of including the quantum effects through $\zeta_{\mathrm{dm}} \neq 0$ is to supply some upward distortions in the plateau portion of $\log \rho_s$, as will be seen in Figure 2 in [98], which, however, hardly leaves noticeable effects on the magnified view around the present epoch.

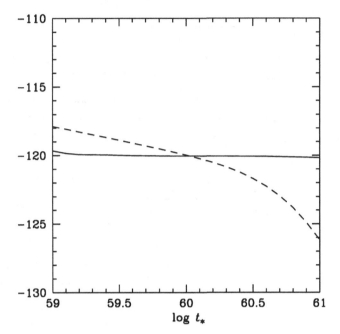

Fig. 5.9. A magnified view of $\log \rho_{\mathrm{s}}$ (solid) and $\log \rho_*$ (dashed) in the lower diagram of Fig. 5.8 around the present epoch. Note that ρ_{s} is very flat in this diagram extending back to the past of $z = 5.2$–6.9 for the assumed age $(1.1$–$1.4) \times 10^{10}$ years.

the mini-inflations occurs around the present epoch, $(1.1$–$1.4) \times 10^{10}$ years, corresponding to $\log t_* = 0.434\tau = 60.1$–$60.2$.

Another goal of adjusting parameters is to keep ρ_{s} lying below ρ_* by several orders of magnitude for $\log t_* \sim 45$ for the nucleosynthesis and also for $\log t_* \sim 50$ for the decoupling time, in order not to jeopardize the success of the standard cosmology.

The background itself undergoes power-law expansion with the effective exponent $\frac{1}{3}$, $\frac{1}{2}$ and $\frac{2}{3}$ corresponding to dominances by the kinetic term of the scalar fields, radiation, and dust, respectively, as one can see in the middle display.

Almost synchronized with the mini-inflation in the scale factor, the density of the scalar fields, ρ_{s}, overtakes the matter density, ρ_*, as shown in the bottom display. It is thus obvious that the plateau of ρ_{s} mimics Λ with $\Omega_\Lambda \simeq 0.5$, as expected. This can be seen more clearly in Fig. 5.9, in which the crossing of the two densities is shown around the present epoch on a magnified scale.

Again on the original scale in the bottom display of Fig. 5.8, we find that both ρ_{s} and ρ_* fall off like t_*^{-2} as an overall behavior, but "interlacing"

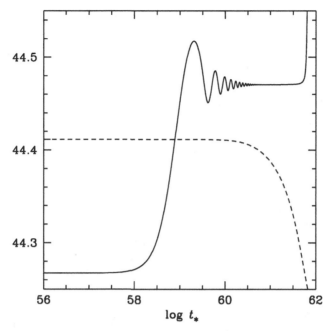

Fig. 5.10. A magnified view of σ (solid) and $0.02\Phi + 44.25$ (dashed) in the upper diagram of Fig. 5.8. Note that the vertical scale has been expanded by a factor of approximately 330 compared with Fig. 5.8. With the time variable $\tau = \ln t_*$, the potential V grows as the multiplying factor $t_*^2 = e^{2\tau}$. The potential wall for σ becomes increasingly steep, thus confining σ further toward the bottom of the potential, which is manifested by an oscillatory behavior of σ with ever-increasing frequency measured in terms of τ. The growing V causes Φ_* eventually to fall downward, resulting in the collapse of the confining potential wall. The stored energy is then released to unleash σ.

each other. A tracking behavior is obviously seen. This implies that we successfully inherit the scenario of a decaying cosmological constant, as we planned. The overtaking mentioned above is followed by a period during which ρ_s tucks under ρ_*, thus returning to the usual power-law expansion of the universe.

Each mini-inflation is associated with a sudden rise of σ accompanied by a simultaneous flip of χ. Except for these pulsations, both scalar fields stay dormant, a hesitation behavior.

Figure 5.9 is a magnified view of ρ_s and ρ_* near the present epoch, while Fig. 5.10 shows σ and χ around the present epoch in the scale magnified by a factor of 330 relative to the top display of Fig. 5.8. Most noticeable is the fine oscillation of σ, through which the "heart" of the underlying mechanism manifests itself, as will be explained briefly.

The scalar fields came to a stop simply because of the cosmological friction. For this reason the places where they spent their dormant period have nothing to do with potential minima. As time elapses, however, t_*^2 in front of the potential on the right-hand sides of (5.60) and (5.61) keeps growing until it offsets the smallness of the factor $e^{-4\zeta\sigma}$. The critical time may be given by

$$2\tau - 4\zeta\sigma \sim 0, \qquad (5.67)$$

where σ here should be substituted from the value of σ during the dormant phase. It may happen that σ then finds itself within a "basin of attraction" due to the force from the oscillating potential with the wall proportional to χ^2, as one sees from Fig. 5.8. Consequently, σ is trapped near one of the minima of the sinusoidal potential, as represented by the oscillation which is so tiny in magnitude that one detects it only after sufficient magnification of the plot.

This trapping, becoming ever more stringent, lasts until χ also starts falling finally toward the minimum at $\chi = 0$. This causes a sudden collapse of the barrier that has kept σ from being lured prematurely to the fly-out course. The energy accumulated until then is now released to catapult σ.

The above analysis suggests at the same time that the location of σ at the end of the dormant phase may be off the basin of attraction of the sinusoidal potential. If this is the case, σ is no longer trapped, resulting in a transition to the asymptotic phase, in which σ obeys essentially (5.67), now with σ re-interpreted as a function of τ. We notice that (5.67) is the same as (4.100). Notice also that χ moves toward the asymptotic value $\chi = 0$ so slowly that one can hardly recognize the change in the limited range of τ.

An example of this type of solution is illustrated in Figs. 5.11–5.13, corresponding to Figs. 5.8–5.10, respectively.

Quite surprisingly, the two solutions exhibiting qualitative differences emerge on changing the initial value σ_1 only slightly. We next move on to discuss how this sensitive dependence comes about.

For this purpose, we show Fig. 5.14 in which the behavior of $\kappa\sigma/(2\pi)$, the phase of the sine function in (5.58), for σ_1 that starts at 6.7544, gives a mini-inflation around the present epoch, sweeping at an increment 0.0022 up to 6.7764. We are going to trace step by step how a change of σ_1 within a small interval yields results that are so different from each other.

Look at the first curve 7544, corresponding to $\sigma_1 = 6.7544$. We see that σ starts slowing down, reaching a plateau of the curve. When σ slowly starts again to move forward around $\log t_* \sim 59$, it finds itself almost

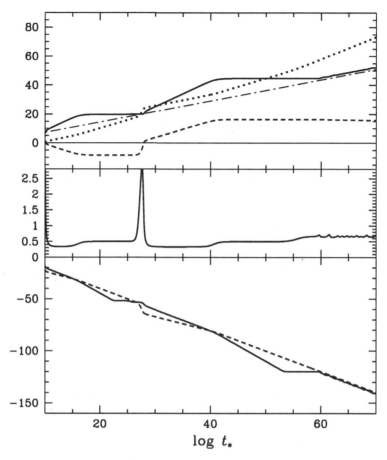

Fig. 5.11. An example of the solution exhibiting no mini-inflation around the present epoch, though another mini-inflation at $\log t_* \sim 27$ is still present. Symbols and initial values are the same as those explained in Fig. 5.8, except that σ_1 is slightly different, $\sigma_1 = 6.761$.

passing the middle of the downward slope of $\sin(\kappa\sigma)$ that is coming up from far below due to the factor t_*^2, therefore being well inside the "basin of attraction." It starts to fall toward the minimum at $\kappa\sigma \approx (3/4)2\pi$ modulo 2π, though the exact location of the minimum deviates somewhat from that of a sine curve due to the presence of other terms in the potential $V(\sigma)$. It approaches an oscillating phase, which was shown in the enlarged plot of Fig. 5.10. Complete trapping continues until the barrier suddenly collapses, resulting in a catapult-like ejection.

The next curve, labeled 7566, with a slightly larger $\sigma_1 = 6.7566$ reaches a correspondingly higher plateau. The amplitude of the oscillation appears

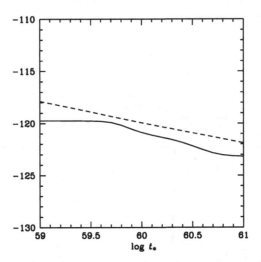

Fig. 5.12. A magnified view of $\log \rho_s$ (solid) and $\log \rho_*$ (dashed) in the lower diagram of Fig. 5.11 around the present epoch. Unlike in the corresponding plot Fig. 5.9, ρ_s fails to cross ρ_*. The maximum of Ω_Λ in this plot is only ≈ 0.2.

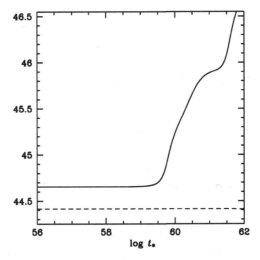

Fig. 5.13. A magnified view of σ (solid) and $0.02\Phi + 44.25$ (dashed) in the upper diagram of Fig. 5.11. Unlike in the plot of Fig. 5.10, σ behaves much more smoothly, undergoing no oscillation.

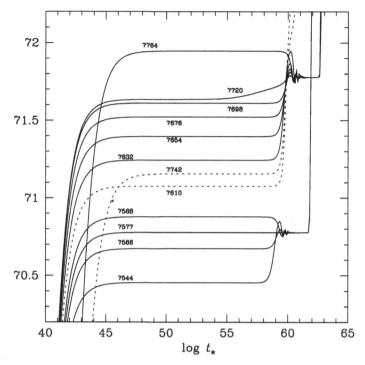

Fig. 5.14. Another magnified view of the behavior of σ, expressed in terms of $\kappa\sigma/(2\pi)$, the phase of the sinusoidal portion of the potential, for the initial value σ_1 varying from 6.7544 through 6.7764 with an increment of 0.0022, with an added exception for 7577, exhibiting a straight fall-off without any oscillation. The last four digits are attached to each curve. Dotted curves are for the escape-type solution.

to be almost half that of the previous curve because σ of the plateau happens to be closer to the potential minimum. We added the curve 7577 with an exceptional increment of 0.0011 of the initial value, in order to demonstrate that σ falls straight to the minimum without any noticeable oscillation.

The fourth curve, labeled 7588, appears to find itself having slightly "overshot" the minimum, but is brought back to the position of the minimum.

With the fifth curve, labeled 7610, a dotted curve, we find something different. The overshooting is such that σ finds itself exceeding the maximum of the potential near $\kappa\sigma = 2\pi \times 71.25$, and thus escapes the basin of attraction. Without being trapped, σ flies out to make a "premature transition" to the asymptotic state, as we saw in the model of a simple hesitation with only one scalar field. There is no mini-inflation, as was shown on a larger scale in Figs. 5.11–5.13.

The next curve in our trial is that labeled 7632, for which σ is trapped in the neighboring basin of attraction. This time it experiences a rather large amount of overshooting, but is somehow pulled back, settling down to the next minimum near $\kappa\sigma = 2\pi \times 71.75$. The following three curves show more or less the same features, but with smaller spacings separating neighboring plateaus. The curve 7720 shows signs of a change, leading to the drastically different curve 7742. The position of the plateau goes down rather abruptly, coming close to that of 7610, and, for that reason, we encounter again a fly-out course without a mini-inflation. The behavior is similar to that of 7610, illustrated in Figs. 5.11–5.13. Looking at these examples, we have obviously a phenomenon of repetition.

The abrupt fall of the plateau position is due to the change in what happened during the first mini-inflation. After the curve 7742, the plateau position resumes a rising trend, as seen for 7764, apparently going past the minimum near $2\pi \times 71.75$.

In these examples of the escape course we evidently have a process of exceeding a potential maximum. It is as if we were walking near a watershed of gentle slope. Before reaching the ridge, our course is rather continuous against the change of the initial value σ_1, but a slight difference in position where we come close to the maximum results in very different futures; flying out or being trapped by a neighboring minimum. One may imagine how a golf ball falls or misses the target hole depending sensitively on how one putts a ball.

In this way we come to understand the nature of the underlying mechanism behind the very different outcomes arising rather abruptly as the initial value changes slightly and smoothly. We also remind the reader that most chaotic behaviors in nonlinear dynamics come essentially from the way in which a system, simple or complicated, passes a maximum or a saddle point of the potential [101].

We fine-tuned the parameters and initial values *moderately* in order to obtain a significant value for Ω_Λ at the age of around 10^{10} years, for which we find *crossing* of ρ_s and ρ_*. This naturally induces an extra acceleration. In Fig. 5.9, a magnified view of the lower diagram of Fig. 5.8, we present the more detailed behavior of the densities, from which we can calculate Ω_Λ and h_0, the Hubble constant in units of $100\,\mathrm{km\,s^{-1}\,Mpc^{-1}}$, yielding the values listed in Table 5.2. We consider them reasonably good fits to the observation.

Given the strong dependence on the initial values, however, how likely is it that one should reach a solution as this? In this connection we first remind the reader that the chance of coming across solutions of the "trap-and-eject" kind, as in Fig. 5.8, is rather high, because the basin of attraction is very wide, as was shown in the analysis related to Fig. 5.14. This

Table 5.2.   The assumed age t_{*0} in 10^{10} years, $\log t_{*0}$ in the reduced Planckian unit system, and Ω_Λ and h_0 for the solution in Fig. 5.8.

$t_{*0}\,(10^{10}\,\text{years})$	$\log t_{*0}$	Ω_Λ	h_0
1.1	60.11	0.62	0.81
1.2	60.15	0.67	0.77
1.3	60.18	0.72	0.74
1.4	60.22	0.76	0.72

is related mainly to the choice of the initial value of σ. There are other parameters that affect solutions in many different ways. Roughly speaking, however, it would take hours instead of days to find solutions of the desired nature, once we are comfortably in a cruising mode of numerical integration, like NDSolve of Mathematica.

We still have to adjust parameters carefully, particularly to avoid having too large a value of ρ_s for $\log t_* \sim 45$, as the time of primordial nucleosynthesis. This has something to do with the parameter κ that affects strongly the durations of the mini-inflations, for example. Adjustment of this kind would take much more time. We still say that this can be done, essentially within the context of reasonably fine tuning, also with parameters of the order of unity in the reduced Planckian unit system. We add also that we have never attempted systematically to sweep all over the parameter space.

5.4.3 Qualitative features unique to the two-scalar model

Apart from fitting parameters to realistic data rather successfully, we discuss some of the qualitative features which are unique to the model discussed above.

From the example shown in Fig. 5.8, we find that the accelerating universe that we see now is not a "once-and-for-all event" in the whole history of the universe; it is just one of the repeated occurrences. This lessens the seriousness of the coincidence problem, though it never removes it entirely. A similar idea of oscillating occurrence was proposed by Dodelson *et al.* [102], though they do not include the second scalar field, resulting in the situation that σ is eventually trapped in a minimum of the sinusoidal potential, hence entering a final phase of eternal inflation.

This "repeated-occurrence picture" may be contrasted with the one we may derive from the quintessence models, in which we naturally view the acceleration now taking place as lasting forever. Models with negative pressure, which drives eternal expansion, can be shown, however, to result in a future horizon, which provides an obstacle to defining observables in terms of an S-matrix [103], the same difficulty in string theory as the de Sitter universe. The present model might provide a solution to this issue.

We finally add comments on the very sensitive dependence of our solution on the parameters, including initial values, as shown before, which might be interpreted as chaos-like, which seems to present a new viewpoint on the solution of the cosmological equations.

We argued that the sensitive dependence on initial values comes essentially from how one exceeds a maximum or a saddle point of the potential. This was shown particularly for the dependence on σ_1 in connection with Fig. 5.14. This is obviously a feature shared with almost any example of *chaotic* behavior exhibited by many nonlinear systems [101]. The question that then arises is that of whether our solution is authentically chaotic. Our reply, as will be presented below, is "No." Our solution is only "nearly" chaotic.

We can show that orbits in the phase diagram do not reveal anything like a "strange attractor." We encounter instead a fixed-point attractor, though the trajectory exhibits somewhat complicated behavior during an intermediary period. However, its falling finally into a fixed point is obviously due to the fact that there is a cosmological frictional force, corresponding to the asymptotic behavior shown in Fig. 5.8, for example. This fact conceals the underlying chaotic behavior incompletely. In fact one of the present authors (K. M.) revealed the chaotic nature of the cosmological equation including a scalar field if the frictional force is turned off [104, 105]. Even with the frictional force, some features of the chaotic dynamics still manifest themselves partially through the sensitive dependence on initial values.

Another feature signaling the chaotic nature of the solution is a qualitative change that occurs "suddenly," while the parameters vary smoothly. An example has already been shown in Fig. 5.14, where a fly-out solution occurs sporadically, while σ_1 changes continuously. This implies that the number of mini-inflations changes abruptly. It is clear that the real origin is the existence of maxima of the potential, leaving the cosmological environment almost of secondary importance. For this reason we looked closely at the system of equations obtained from our original equations with the cosmological component removed;

$$\sigma'' + \frac{1}{2}\sigma' + \frac{\partial V}{\partial \sigma} = 0, \tag{5.68}$$

$$\chi'' + \frac{1}{2}\chi' + \frac{\partial V}{\partial \chi} = 0, \tag{5.69}$$

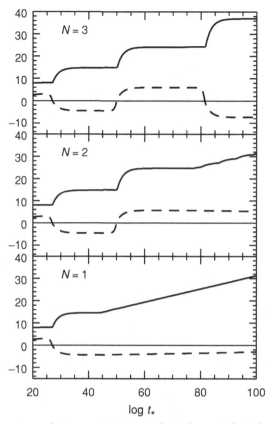

Fig. 5.15. Examples of the solution of (5.68) and (5.69) exhibiting different numbers (N) of jumps. The initial values of σ_1 and σ'_1 are the same as in Fig. 5.8, whereas $\chi_1 = 2.50, 2.59$, and 2.60 for the examples of $N = 1, 2$, and 3, respectively, while $\chi'_1 = 0.0$.

with the same potential as that given by (5.58), which is shown graphically in Fig. 5.7. We chose $b' = \frac{1}{2}$, fixing the coefficient of the frictional term to be a constant.

Some of the typical solutions are shown in Fig. 5.15 for varied initial values χ_1 and χ'_1, while σ_1 and σ'_1 are held fixed. The horizontal axis is for $\tilde{\tau} = \log t_* = \tau/\ln 10$, with $\tau = \ln t_*$ the variable in (5.68) and (5.69).

In the bottom diagram, $\sigma(\tau)$ undergoes a sudden rise (jump) at $\tilde{\tau} \approx 28$. It soon levels off and reaches a straight line after $\tilde{\tau} \approx 44$, representing an asymptotic state. This curve exhibits a single jump, denoted by $N = 1$ in Fig. 5.15. Likewise the middle diagram is an example of the occurrence of two jumps before the system settles into the asymptotic state. Each jump of σ is associated with a flipping behavior of $\chi(\tau)$, which is also a

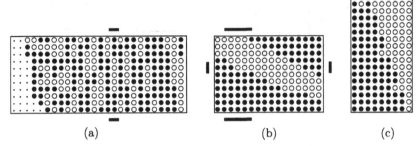

(a)   (b)   (c)

Fig. 5.16. In plot (a), the distribution of the number of "jumps" N, for the initial values $\chi_1 = 2.50$–2.75 shown in the horizontal direction with the interval 0.01, and $\chi'_1 = 0.240$–0.250 in the vertical direction with the increment 0.001, is illustrated by open and filled circles corresponding to $N = 2$ and $N \geq 3$, respectively, while dots are for $N = 1$. It appears as if a filament-like behavior is occurring everywhere, suggesting a fractal structure. If that is the case, it might indicate that the solution is of a chaotic nature. In order to check whether this is indeed the case, we show, in plot (b), a magnified view. The same distribution is shown for the range of χ_1 between 2.640 and 2.655, indicated by the two short bars below and above the horizontal frames in plot (a), magnified by a factor of ten in the horizontal direction, but unchanged in the vertical direction. We find simply smooth boundaries without further fractal nature, no longer filament-like boundaries as apparently occur in plot (a). Plot (c) is a further magnification of a portion marked by horizontal and vertical marks of plot (b); $\chi_1 = 2.641$–2.645 and $\chi'_1 = 0.2452$–0.2468. The magnification ratios compared with the original plot (a) are 62.5 and 6.25 in the χ and χ' directions, respectively. The boundary is still a smooth line. The wiggles observed on some parts of the boundary are probably due to the errors in the calculation itself.

sudden change. The solution in the top diagram has three or more jumps and the same number of flips. Solutions of different N are considered to be qualitatively different from each other. Now a unique feature is that the solution exhibits this qualitative change suddenly for smooth changes of χ_1 and χ'_1, as illustrated in Fig. 5.16, in which the distribution of N is plotted in two-dimensional χ_1–χ'_1 space.

In plot (a), where the ranges $\chi_1 = 2.50$–2.75 and $\chi'_1 = 0.24$–0.25 are plotted in the horizontal and vertical directions, respectively, we recognize a filament-like structure of the regions for each value of N. Plot (b) provides a magnified view of a portion of (a) indicated by the short bars below and above the horizontal frames but with the same vertical range.

The occurrence of rather smooth boundaries suggests that the mesh used in plot (a) must have been too coarse. It does not appear that there is fractal structure in this phase space. This nearly rules out the possibility that the solution is truly chaotic in spite of the sensitive dependence on initial values.

In plot (c), we further magnified a small portion indicated by two pairs of short bars marked outside the frames in (b). We reinforce the above conclusion of the absence of fractal structure. The small wiggles on the boundary seem to be due to the limited accuracy of the calculation.

We admit that we need to perform a more careful analysis, particularly regarding the accuracy of the calculation, before drawing a final conclusion. We nevertheless point out that there is a difference from the solution due to Maeda and Easther [104], who discovered a fully chaotic behavior of the solution of the two-scalar equation, but with the frictional force dropped. With certain reservations for the moment, we try to discuss how a novel view is going to emerge if the solution is in fact very sensitive to the choice of initial values.

First, it may suggest a possible new form of "dissipative structures" in the sense due to Prigogine [106], which might not be purely mathematical. Any coastal line on the real map is not purely fractal down to infinite order, as one can generate by using Koch's function. A very sensitive dependence on initial values may downgrade the predictive power of the theory. We remind the reader, however, that the rather low performance of weather-forecasting systems, predicting what are known to be examples of chaotic phenomena in nature, never prevents us from thinking with conviction that the way in which weather evolves is based on underlying physical processes. We know many similar examples in nature. Why is the universe an exception? This might be the price of solving the problem of the cosmological constant [107].

According to a traditional view we should try to find solutions that depend on initial values as little as possible. This belief is connected also with the expectation that we live in an asymptotic state that can be reached from almost any initial values, namely there is an attractor. This attitude has been adopted widely, and was emphasized particularly by Damour and Polyakov [13], as well as by ourselves when the quintessence scenario was discussed. According to our solution in this subsection, however, we are right in the midst of a transient state. Living with Λ might force us to accept a change.

We add that the example, particularly in the top diagram, in Fig. 5.15 appears to show that there is a "periodicity" with respect to the variable τ. It is interesting to note that a model of game dynamics with more than three species also generates near periodicity with respect to $\ln t$, though it is not clear whether it has anything in common with our model of the underlying mechanism [108].

Obviously friction is crucially important in bringing about oscillation, and the resulting "period" depends more on the initial values than on the parameters prepared in the Lagrangian. In this respect the behavior of our solution is similar to what is known as "relaxation oscillation," sharing something in common even with such natural "periodic" phenomena as patterns on the surface of sand dunes, bumpy ski slopes, and seashells [109], which are considered to be highly nonlinear effects. We are likely looking at repeated phenomena quite different from what has been discussed basically in connection with linear harmonic oscillators like the mechanism studied in [110], for example.

6

Quantum effects

In the previous chapter we discussed the revised model of the scalar–tensor theory, a scale-invariant model, which is expected to provide a better understanding of the extra acceleration of the universe, indicating the existence of a small but nonzero cosmological constant. We stayed, however, mainly within the confines of the classical theory, though we have discussed already how important the quantum corrections due to the interaction among matter fields could be. In this chapter we develop quantum theory related to breaking of scale invariance, in particular through the effect of an anomaly.

We emphasize that the quantum theory we are going to discuss applies to matter fields in flat space-time, which is more conventional than quantum effects of a scalar field in curved space-time, or even quantization of space-time itself. The gravitational scalar field will remain classical on most occasions except for discussing its self-mass.

Also re-iterated is that the transformation $ds^2 \rightarrow \Omega^2(x)\, ds^2$, or $g_{\mu\nu} \rightarrow \Omega^2(x)g_{\mu\nu}$, which we have discussed since Chapter 3 under the nomenclature "conformal transformation," or "Weyl rescaling," is a *local* change of the space-time distance, not to be confused with a *global coordinate transformation* in flat space-time, $x^\mu \rightarrow (x^\mu + \lambda^\mu x^2)/(1 + \lambda x + \lambda^2 x^2)$ with four parameters λ_μ, which is sometimes called a "special conformal transformation." Likewise, the transformation which we call a "scale transformation," for which we sometimes use the terminology "dilatation," is a *global version* of our conformal transformation, but is not the same as the global coordinate transformation $x^\mu \rightarrow \lambda x^\mu$. In spite of this difference in principle, the two transformations may lead to the same conclusion, in some applications.

We begin sections 6.1 and 6.2 with the classical argument concerning the dilatation current associated with scale invariance, because it will help us later to understand spontaneously broken scale invariance. Throughout this chapter we give an explicit formulation of a scalar field Φ as a representative of matter fields, as we did in the preceding chapters. We have now, however, another reason for this choice; we know no easy way to define the dilatation current for a spinor field, though we expect that the final results obtained from the model of a scalar field will remain unaffected in general. In section 6.2 we focus on preparing the formulation in D dimensions and going to the E frame, thus realizing a spontaneous breakdown of dilatation invariance.

In section 6.3 we develop a calculation of one-loop integrals, first assuming the self-coupling of a real scalar field and then quantum electrodynamics (QED) of a complex scalar field. We show explicitly how a pole $(D-4)^{-1}$ is canceled out by a factor $(D-4)$, yielding a nonzero and finite result, an anomaly.

One of the main results of this chapter is that the scalar field σ does play the role of a Nambu–Goldstone (NG) boson in a rigorous sense, a dilaton, but becomes a pseudo-NG boson when the quantum effect is introduced, and re-appears as the non-Newtonian field featuring a nonzero mass, namely a finite force-range and violation of the WEP.

Non-Newtonian gravity is then discussed in section 6.4, by extending the scalar QED to quantum chromodynamics (QCD). The strength turns out to be much weaker than would naively have been expected, mainly due to the relatively small portion of the nucleon mass arising from the quark mass. This allows us to expect the effect to be discovered barely below the existing upper bounds, in view of the fact that we still have uncertainties in the calculation. We stress intuitive and semi-quantitative aspects, leaving details of the phenomena to the review articles [40, 111].

As another outcome of quantum effects, we show in section 6.5 that the possible violation of the WEP might be tested experimentally through measurements of the electromagnetic interactions of nuclei, if the scalar field couples to the classical Maxwell Lagrangian in the J frame, against the assumed scale invariance.

Section 6.6 is devoted to discussing the relevance of quantum effects to a possible time-variation of the fine-structure constant. The upper bounds obtained so far and the reported "evidence" for the variation [18] will impose a crucial constraint on the theory. In this way we find that a number of phenomena of recent interest are related to each other as manifestations of the dynamics of a gravitational scalar field.

One might raise the question of why we include quantum effects in the E frame rather than in the J frame. As we admit, we do not do this on the basis of a first principle of any kind. There seems to be no reason why we should not do the same in the J frame. We are not quite sure whether the two procedures, conformal transformation and quantization, are mutually commutable. We simply expect that we have developed a model according to which the E frame is closer to the real world than is the J frame, and is qualified to provide a better classical theory as a basis for quantization. From this point of view, the "theoretical" steps through which scale invariance is broken, as will be shown, seem to allow us to obtain a reasonable interpretation by applying quantization at the last stage.

6.1 Scale invariance

As we emphasized before, the models having a nonminimal coupling in the simplest form possess *scale invariance*, another attractive aspect that deserves further attention. This is to be contrasted with the well-known fact that the conventional Einstein–Hilbert term is multiplied by a dimensional constant M_P^2 that obviously breaks scale invariance. This conclusion about scale non-invariance follows also from the matter part, in which the mass terms of the fields introduce (length or time) scales into the theory. We show, however, that the breaking of this invariance can be spontaneous, thus providing a new approach to understanding the origin of particle masses in unified theories.

We warn the reader here that the Λ term in the J frame is outside of this idea. We must have some rationale for proposing a principle that applies only partially.

We now start with the Lagrangian (2.1),

$$\mathcal{L} = \sqrt{-g}\left(\tfrac{1}{2}\xi\phi^2 R - \tfrac{1}{2}\epsilon g^{\mu\nu}\,\partial_\mu\phi\,\partial_\nu\phi + L_{\text{matter}}\right), \tag{6.1}$$

but with the matter Lagrangian given by

$$L_{\text{matter}} = -\tfrac{1}{2}g^{\mu\nu}\,\partial_\mu\Phi\,\partial_\nu\Phi - \frac{f}{4}\phi^2\Phi^2 - \frac{\lambda_\Phi}{4!}\Phi^4. \tag{6.2}$$

The last term represents a self-coupling of Φ with a coupling constant λ_Φ which should be roughly of the order of unity. Recall that Φ is a representative of matter fields, as in section 3.3, and its interaction ought to be much stronger than the gravitational interaction. One should notice that, according to our revised model discussed in subsection 4.4.3 of the previous chapter, the second term on the right-hand side of (6.2) shows that ϕ (the gravitational scalar field) couples to the matter *directly* at the level of the Lagrangian, thus violating one of the basic premises of the prototype BD model. It then follows that the WEP is violated, as expected for the fifth force. This might be viewed as a challenge to what Dicke had taken for granted experimentally. Also we assume at this level that Φ is massless; no term like $(m^2/2)\Phi^2$ has been introduced, which is another important difference from the prototype BD model. Notice also that the second term on the right-hand side of (6.2) is slightly different from (4.129), because we want here to follow specifically the way a mass arises in the Higgs mechanism.

It should also be noted that we included only quartic terms in the sector of the matter scalar field, except for the kinetic term. This implies that all the constant multipliers are *dimensionless*. This is due to the fact that any bosonic fields have the dimension of mass, as is determined by requiring that the kinetic terms should have the dimension of $(\text{mass})^4$. It also follows that ξ in the nonminimal coupling is again dimensionless; hence there is no dimensional coupling constant in this theory, as long as we confine ourselves to the simplest form of nonminimal coupling term, as in the prototype BD model. Scale invariance is hence a unique feature of this model. We are going to pursue its consequences.

One may allow terms like $\phi^4, \phi^3\Phi$, and $\phi\Phi^3$, which are multiplied by dimensionless coupling constants. We consider only terms of crucial importance, though terms of odd powers of Φ can be excluded by imposing certain symmetries of strong interactions.

In passing we note that, in exploiting scale invariance with its spontaneous breaking, we share the same basic approach with Wetterich [112], though with considerable differences in details.

Let us derive the field equations. "Einstein's equation" is of the same form as (2.6):

$$2\varphi G_{\mu\nu} = T_{\mu\nu} + T^\phi_{\mu\nu} - 2(g_{\mu\nu}\,\Box - \nabla_\mu\nabla_\nu)\varphi, \qquad (6.3)$$

but with the different $T_{\mu\nu}$. Equations (6.1) and (6.2) give

$$T_{\mu\nu} = \partial_\mu\Phi\,\partial_\nu\Phi + g_{\mu\nu}\left(-\frac{1}{2}g^{\rho\sigma}\,\partial_\rho\Phi\,\partial_\sigma\Phi - \frac{f}{4}\phi^2\Phi^2 - \frac{\lambda_\Phi}{4!}\Phi^4\right), \qquad (6.4)$$

from which it also follows that

$$T = -(\partial\Phi)^2 - f\phi^2\Phi^2 - \frac{\lambda_\Phi}{3!}\Phi^4. \qquad (6.5)$$

Notice that T of a scalar field fails to vanish even without mass.

Corresponding to the ϕ equation (2.13), we obtain

$$\xi\phi R + \epsilon\,\Box\phi - \frac{f}{2}\phi\Phi^2 = 0. \qquad (6.6)$$

In the same way we obtain

$$\Box\Phi - \frac{f}{2}\phi^2\Phi - \frac{\lambda_\Phi}{3!}\Phi^3 = 0. \qquad (6.7)$$

Multiplying (6.6) by ϕ yields

$$2\varphi R + \epsilon\phi\,\Box\phi - \frac{f}{2}\phi^2\Phi^2 = 0. \tag{6.8}$$

Taking the trace of (6.3), and using (6.5), we derive

$$-2\varphi R = -\epsilon\,(\partial\phi)^2 - 6\,\Box\varphi - (\partial\Phi)^2 - f\phi^2\Phi^2 - \frac{\lambda_\Phi}{3!}\Phi^4. \tag{6.9}$$

By eliminating R from (6.8) and (6.9) we obtain

$$\left(6 + \epsilon\xi^{-1}\right)\Box\varphi = -\tfrac{1}{2}\,\Box\Phi^2, \tag{6.10}$$

which may be put into the form

$$\Box\left(\varphi + \tfrac{1}{2}\zeta^2\Phi^2\right) = 0. \tag{6.11}$$

Since ϕ has direct coupling to matter, (2.8) no longer holds exactly.

Now let us go into a more detailed discussion of scale invariance. As in Chapter 3, we introduce the new metric $g_{*\mu\nu}$ defined by

$$g_{\mu\nu} = \Omega^{-2}g_{*\mu\nu}, \tag{6.12}$$

but, unlike in Chapter 3, Ω is now a *constant* that is independent of x. In this sense we are considering a transformation

$$g_{\mu\nu} \to g_{*\mu\nu} = \Omega^2 g_{\mu\nu}, \tag{6.13}$$

applied to the field $g_{\mu\nu}$. It is convenient to consider an infinitesimal transformation given by

$$\Omega = e^\Lambda \qquad (|\Lambda| \ll 1), \tag{6.14}$$

in terms of which (6.13) is expressed as

$$g_{\mu\nu} \to g_{\mu\nu} + \delta g_{\mu\nu}, \tag{6.15}$$

where

$$\delta g_{\mu\nu} = 2\Lambda g_{\mu\nu}. \tag{6.16}$$

On the other hand, forcing Ω in (3.14) to be constant yields

$$R = \Omega^2 R_*. \tag{6.17}$$

Equations (3.7) and (3.8) remain unchanged:

$$g^{\mu\nu} = \Omega^2 g_*^{\mu\nu}, \tag{6.18}$$

$$\sqrt{-g} = \Omega^{-4}\sqrt{-g_*}. \tag{6.19}$$

We then find that, on applying the transformation

$$\phi = \Omega \phi_* \tag{6.20}$$

to ϕ, the nonminimal coupling remains the same as before:

$$\tfrac{1}{2}\sqrt{-g}\xi\phi^2 R = \tfrac{1}{2}\sqrt{-g_*}\xi\phi_*^2 R_*, \tag{6.21}$$

implying that this term is *scale-invariant*.

Equation (6.20) can also be put into the form of an infinitesimal transformation:

$$\phi \to \phi + \delta\phi, \tag{6.22}$$

where

$$\delta\phi = -\Lambda\phi. \tag{6.23}$$

We find immediately that the kinetic term is then also scale-invariant.

The same is true also for Φ. In fact, the kinetic term as well as the interaction terms (multiplied by $\sqrt{-g}$) remain invariant under

$$\Phi = \Omega\Phi_*, \tag{6.24}$$

with the infinitesimal version

$$\delta\Phi = -\Lambda\Phi. \tag{6.25}$$

With an invariance under a global transformation, we can derive a conserved current thanks to *Noether's theorem*:

$$\Lambda J^\mu = \sum_{u=g_{\mu\nu},\phi,\Phi} \frac{\partial \mathcal{L}}{\partial \partial_\mu u} \delta u, \tag{6.26}$$

where the Lagrangian is assumed to contain only up to the first derivatives of the fields. For this reason, we must remove the second derivatives of $g_{\mu\nu}$ in the way described in Appendix C. Leaving the details again to Appendix M, we give the result

$$J^\mu = \tfrac{1}{2}\sqrt{-g}g^{\mu\nu}\,\partial_\nu\left(\xi\zeta^{-2}\phi^2 + \Phi^2\right). \tag{6.27}$$

Its conservation law is written as

$$\partial_\mu J^\mu = \tfrac{1}{2}\sqrt{-g}\,\Box\left(\xi\zeta^{-2}\phi^2 + \Phi^2\right), \tag{6.28}$$

where the right-hand side vanishes thanks to (6.11), verifying that J^μ is conserved.

One might be curious about why no derivative of the metric appears in J^μ. However, we find the derivative of ϕ which is mixed with $g_{\mu\nu}$, thus allowing J_0 to serve as a generator of the transformation of the metric as well.

6.2 The dilaton as a Nambu–Goldstone boson

So far, the field as represented by Φ has no mass, and hence is lacking a crucial aspect of reality. We are going to show, in this connection, that the conformal transformation discussed in Chapter 3 gives an important clue. To further develop the idea, however, we are going to rely on the concept of an *anomaly* known in quantum field theory. In this process, the technique of *regularization* is indispensable. We specifically rely on the *method of continuous dimensions*. To fit this approach we will write the following equations in *arbitrary* space-time dimensions.

Let the space-time dimension be denoted by D. One may start with positive integers, but can easily extend the result to continuous numbers, going finally to the limit $D \rightarrow 4$. We also use the symbol

$$d \equiv D/2. \tag{6.29}$$

We now move to the E frame by applying the local conformal transformations (3.6) and (3.7):

$$g_{\mu\nu} = \Omega^{-2} g_{*\mu\nu}, \quad \text{or} \quad g^{\mu\nu} = \Omega^2 g_*^{\mu\nu}, \tag{6.30}$$

with a space-time-dependent $\Omega(x)$. Equation (3.8) is then replaced by

$$\sqrt{-g} = \Omega^{-D} \sqrt{-g_*}. \tag{6.31}$$

The transformation rule of R in D dimensions is given in Appendix G. As will be shown later, however, a significant result emerges only if $D - 4$ is multiplied by a divergent integral (represented by a pole at $D = 4$). We find that we encounter no such situation with R itself. For this reason we may use (3.14) as it stands:

$$R = \Omega^2 (R_* + 6 \,\Box_* f - 6 g_*^{\mu\nu} f_\mu f_\nu), \tag{6.32}$$

where

$$f = \ln \Omega \quad \text{and} \quad f_\mu = \partial_\mu f. \tag{6.33}$$

Notice that the exponent of Ω in (6.32) is always 2 for any D. For this reason, the exponent of Ω in (3.19) is in fact $2 - D$ for $D \neq 4$. It then follows that the exponent of Ω in (3.20) is now replaced by $2 - D$:

$$\xi\phi^2 = \Omega^{D-2}. \tag{6.34}$$

The relation between ϕ and σ remains essentially the same as before,

$$\phi = \xi^{-1/2} e^{\zeta_D \sigma}, \tag{6.35}$$

but with ζ_D now given by

$$\zeta_D^{-2} = 4\frac{D-1}{D-2} + \epsilon\xi^{-1}, \tag{6.36}$$

as shown in Appendix M in more detail. It is clear that ζ_D defined here does not exhibit any special behavior for $D \to 4$, justifying the use of the previous value at $D = 4$, namely (1.14):

$$\zeta^{-2} = 6 + \epsilon\xi^{-1}. \tag{6.37}$$

We also have

$$\Omega = \exp\left(\frac{1}{d-1}\zeta_D\sigma\right), \tag{6.38}$$

which replaces (3.40).

Now we move on to the portion L_{matter}. Consider first the kinetic term of Φ. In order to extend (3.59), we first note that

$$-\tfrac{1}{2}\sqrt{-g}g^{\mu\nu} = -\tfrac{1}{2}\sqrt{-g_*}g_*^{\mu\nu}\Omega^{-D+2}, \tag{6.39}$$

on which basis we define Φ_* by

$$\Phi = \Omega^{d-1}\Phi_*, \tag{6.40}$$

making the kinetic term canonical:

$$-\tfrac{1}{2}\sqrt{-g}g^{\mu\nu}\,\partial_\mu\Phi\,\partial_\nu\Phi = -\tfrac{1}{2}\sqrt{-g_*}g_*^{\mu\nu}\,\mathcal{D}_\mu\Phi_*\,\mathcal{D}_\nu\Phi_*, \tag{6.41}$$

where

$$\mathcal{D}_\mu\Phi_* \equiv [\partial_\mu + \zeta(\partial_\mu\sigma)]\Phi_*. \tag{6.42}$$

In this way we now have (6.1) in the form

$$\mathcal{L} = \sqrt{-g_*}\left(\tfrac{1}{2}R_* - \tfrac{1}{2}g_*^{\mu\nu}\,\partial_\mu\sigma\,\partial_\nu\sigma + L_{*\text{matter}}\right), \tag{6.43}$$

where

$$L_{*\text{matter}} = -\tfrac{1}{2}g_*^{\mu\nu}\,\mathcal{D}_\mu\Phi_*\,\mathcal{D}_\nu\Phi_*$$
$$- \exp\left(2\frac{d-2}{d-1}\zeta\sigma\right)\left(\xi^{-1}\frac{f}{4}M_{\text{P}}^2\Phi_*^2 + \frac{\lambda_\Phi}{4!}\Phi_*^4\right). \tag{6.44}$$

We have used (6.35), (6.38), and (4.80). We also indicated M_{P} explicitly to emphasize that a dimensional constant has appeared. We now have the mass term given in terms of the gravitational constant.

It might be useful to summarize the scenario briefly. In this chapter we started with a theory having no dimensional constants: the world without any meter-stick or clock. In the process of conformal transformation, however, we introduced the Planck mass M_P. Notice that M_P^{D-2} is multiplied by the right-hand side of (6.34). No equation of this type could have been written without admitting a dimensional constant because ϕ has a dimension. Allowing a constant that we had excluded from the Lagrangian is precisely the way spontaneous breakdown of a symmetry is formulated.

In many examples of spontaneous violation of symmetry, we think of a potential that induces the breakdown. The same procedure cannot be followed, however, in this case of scale invariance. Nevertheless, the very existence of a solution involving dimensional constants justifies the claim that this is essentially the process of spontaneous breaking of symmetry. A similar example may be found in KK theory, in which part of multidimensional space-time is compactified to a smaller closed space. Spontaneous compactification, as it is called, is based on the fact that Einstein's equation in higher dimensions does allow a solution of the type mentioned.

A possible problem arising from the lack of the potential is that one cannot single out a solution based on a stability argument. In our approach, however, one might expect a different interpretation from the cosmological point of view. Leaving this issue aside for the moment, let us go on to discuss consequences of spontaneously broken scale invariance.

In (6.44) we notice that σ without a derivative appears exclusively in the exponential factor that multiplies the polynomial of Φ_*. We also notice that σ disappears for $d = 2$. In other words, σ has no coupling to matter. The derivative coupling of σ is still present in (6.42). Since we are mainly interested in the situation in which the 4-momentum carried by σ is negligibly small, however, we may conclude from scale invariance that σ decouples from matter.

If we go to the limit $d \to 2$ *naively* in (6.44), σ is totally absent. For this reason we may apply the same recipe as with the Higgs field in the standard theory.

If we assume that

$$f < 0 \quad \text{and} \quad \lambda_\Phi > 0, \tag{6.45}$$

we find $\Phi_* = 0$ to be unstable. On introducing the vacuum expectation value v_Φ and the fluctuating field $\tilde{\Phi}$ according to

$$\Phi_* = v_\Phi + \tilde{\Phi}, \tag{6.46}$$

we obtain for the portion with the exponential in (6.44) set to 1

$$-L_0' = L_{\text{vac}} + \frac{1}{2}m^2\tilde{\Phi}^2 + \frac{1}{2}\sqrt{\frac{\lambda_\Phi}{3}}m\tilde{\Phi}^3 + \frac{\lambda_\Phi}{4!}\tilde{\Phi}^4, \tag{6.47}$$

where

$$v_\Phi^2 = -\frac{3f}{\xi \lambda_\Phi} M_P^2, \tag{6.48}$$

and

$$m^2 = -\frac{f}{\xi} M_P^2. \tag{6.49}$$

Also

$$L_{\text{vac}} = -\frac{3}{8} \frac{f^2}{\xi^2 \lambda_\Phi} M_P^4 \tag{6.50}$$

plays the role of a cosmological constant.

In the classical theory, we put $D = 4$, or $d = 2$, at this stage. Then σ, at least in the static and the long-wavelength limit, is completely decoupled from $\tilde{\Phi}$. Unlike in the prototype BD model, we have no way to observe σ by measuring forces among matter objects. In addition to the fact that σ may still manifest itself in cosmological settings, we may have a different kind of chance to "see" it directly if quantum effects are taken into account, as we will show shortly. Before going into the details, however, we must establish that σ plays the role of a *NG boson*.

Most crucial is the Noether current J^μ together with its conservation. The current which we are going to call the *dilatation current* is now given in terms of $g_{*\mu\nu}$, σ, and Φ_* in the E frame:

$$J^\mu = \tfrac{1}{2}\sqrt{-g_*}\, g_*^{\mu\nu} \left[2\zeta^{-1} \partial_\nu \sigma + (\partial_\nu + 2\zeta(\partial_\nu \sigma)) \Phi_*^2 \right], \tag{6.51}$$

as one can easily derive by substituting (3.7), (3.8), (3.38), and (3.40) into (6.27). Notice that we have a term linear in σ on the right-hand side. This is in fact a signal for spontaneous breaking of symmetry to occur, with the spin-zero field entitled to be called a NG boson. In our particular case of dilatation symmetry, we call it a *dilaton*.

One may form a generator by putting

$$Q = \int d^3x\, J^0, \tag{6.52}$$

which, due to the presence of the linear term, fails to annihilate the vacuum,

$$Q|0\rangle \neq 0, \tag{6.53}$$

another manifestation of spontaneously broken symmetry.

As is also obvious from (6.43), there is no mass term of σ; the dilaton is massless, signaling again that it is responsible for the spontaneous nature of the symmetry breaking. It thus follows that, in the theory in the E frame, Φ_* acquires a nonzero mass m according to (4.92), leaving the dilatation current still conserved.

In order to show that (6.51) is conserved, we must use the equation for σ in the E frame, which will be derived from (6.43) and (6.44):

$$\left(1 + \zeta^2 \Phi_*^2\right) \Box_* \sigma + \tfrac{1}{2}\zeta \, \Box_* \Phi_*^2 + \zeta^2 g^{\mu\nu} \, \partial_\mu \sigma \, \partial_\nu \Phi_*^2 = 0. \tag{6.54}$$

6.3 The contribution from loops

We now depart from the classical theory, extending our consideration to the quantum theory. We will confine our interest to the effect of loops of Φ in the sense of relativistic quantum field theory. Particular emphasis will be placed on the terms which go away as $D - 4$, but are multiplied by poles $(D - 4)^{-1}$ as a representation of divergences of loop integrals. This implies that the terms mentioned above vanish classically, but survive with finite nonzero contributions after quantum effects have been included. This is precisely what we have encountered as quantum anomalies [70].

6.3.1 The coupling to matter of σ, a self-coupled scalar field

We first consider the coupling of σ to Φ_* or $\tilde{\Phi}$, which vanishes classically in the limit $D \to 4$, but gives rise to a finite nonzero coupling as an anomaly. We begin by extracting terms linear in σ from (6.44). This will be done first using an expansion,

$$\exp\left(2\frac{d-2}{d-1}\zeta\sigma\right) \approx 1 + 2\zeta(d-2)\sigma, \tag{6.55}$$

in (6.44) and (4.90), with L_{vac} dropped, giving

$$-L_1' = 2\zeta(\nu - 2)\left(\frac{1}{2}m^2\tilde{\Phi}^2 + \frac{1}{2}\sqrt{\frac{\lambda_\Phi}{3}}m\tilde{\Phi}^3 + \frac{\lambda_\Phi}{4!}\tilde{\Phi}^4\right)\sigma. \tag{6.56}$$

We retain $D \neq 4$, namely $d \neq 2$, before we go to the limit $d \to 2$ at the final stage of the calculation. Divergence coming from loops in quantum field theory is manifested as a pole at $d = 2$, which will leave a finite contribution when it is multiplied by $d - 2$. From this point of view, we try to look for one-loop diagrams, which show up in conjunction with the σ coupling, as given by (6.56). We find three diagrams when we restrict ourselves to the terms quadratic in $\tilde{\Phi}$. The diagrams (a), (b), and (c) in Fig. 6.1 come from the terms on the right-hand side of (6.56) in the order in which they appear there.

For convenience, we start with diagram (c). As a contribution to the source of (nonderivative) σ, we obtain

$$\mathcal{J}_c = -i\zeta\lambda_\Phi(D - 4)(2\pi)^{-D}\int d^D k \, \frac{1}{k^2 + m^2}. \tag{6.57}$$

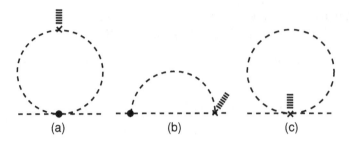

Fig. 6.1. One-loop diagrams yielding the σ (thick, broken line) coupling to $\tilde{\Phi}$ (dashed line). The crosses represent the σ couplings m^2, $\sqrt{3\lambda_\Phi}$, and λ_Φ, all multiplied by $\zeta(D-4)$, for the diagrams (a)–(c), respectively. The filled circles are for the self-couplings of $\tilde{\Phi}$.

This integral exhibits a quadratic divergence in four dimensions, but would be finite if D were sufficiently small ($D < 2$ in this example), and can be analytically continued to larger D. Leaving details to Appendix N, we find, for $d \approx 2$,

$$I_c \equiv \int d^D k \, \frac{1}{k^2 + m^2} = i\pi^2 \left(m^2\right)^{d-1} \Gamma(1-d). \tag{6.58}$$

We substitute this into (6.57), also using

$$(d-2)\,\Gamma\,(1-d) = -\frac{1}{1-d}(2-d)\Gamma(2-d)$$

$$= -\frac{1}{1-d}\Gamma(3-d) = 1. \tag{6.59}$$

In this way we arrive at

$$\mathcal{J}_c = \zeta\lambda_\Phi \frac{m^2}{8\pi^2}, \tag{6.60}$$

giving a finite contribution, as mentioned before; $d-2$ has canceled out the pole at $d = 2$ in $\Gamma(1-d)$.

Let us move on to diagram (a), which gives

$$\mathcal{J}_a = i(2\pi)^{-D}\zeta\lambda_\Phi m^2 (D-4) I_a, \tag{6.61}$$

where

$$I_a = \int d^D k \, \frac{1}{(k^2 + m^2)^2}. \tag{6.62}$$

This integral can be written formally as

$$I_a = -\frac{\partial}{\partial m^2} I_c. \tag{6.63}$$

By using (6.58) in the right-hand side, we obtain

$$I_{\mathrm{a}} = i\pi^2 m^{d-2}(1-d)\Gamma(1-d) = i\pi^2\Gamma(2-d), \tag{6.64}$$

which may further be substituted into (6.61) to yield

$$\mathcal{J}_{\mathrm{a}} = \zeta\lambda_\Phi\frac{m^2}{8\pi^2} = \mathcal{J}_{\mathrm{c}}. \tag{6.65}$$

We finally consider (twice) the diagram (b):

$$\mathcal{J}_{\mathrm{b}} = 12i(2\pi)^{-D}\zeta\lambda_\Phi m^2(d-2)I_{\mathrm{b}}, \tag{6.66}$$

where

$$I_{\mathrm{b}} = \int d^D k\,\frac{1}{[(q-k)^2+m^2](k^2+m^2)}. \tag{6.67}$$

A standard calculation gives

$$I_{\mathrm{b}} = \int_0^1 dx \int d^D k\,\frac{1}{(k^2+\mathcal{M}^2)^2}, \tag{6.68}$$

$$\mathcal{M}^2 = m^2 + x(1-x)q^2, \tag{6.69}$$

from which follows, for $q^2 = -m^2$,

$$\int d^D k\,\frac{1}{(k^2+\mathcal{M}^2)^2} = i\pi^2\Gamma(2-d)\left(m^2\right)^{d-2}\left(1-x+x^2\right)^{d-2}. \tag{6.70}$$

Since $2-d$ is already a factor in (6.66), we may set $d=2$ in the terms other than $\Gamma(2-d)$ in (6.70). In this way we obtain

$$I_{\mathrm{b}} = i\pi^2\Gamma(2-d), \tag{6.71}$$

hence giving

$$\mathcal{J}_{\mathrm{b}} = \frac{3}{4\pi^2}\zeta\lambda_\Phi m^2. \tag{6.72}$$

By adding (6.60), (6.65), and (6.72) together, we finally obtain

$$\mathcal{J} = \frac{1}{\pi^2}\zeta\lambda_\Phi m^2. \tag{6.73}$$

This source term of σ can be represented in terms of an effective Lagrangian:

$$-L_{\Phi\sigma2} = \frac{1}{2\pi^2}\zeta\lambda_\Phi\frac{m^2}{M_{\mathrm{P}}}\tilde{\Phi}^2\sigma, \tag{6.74}$$

Fig. 6.2. The coupling of σ (thick, broken line) to ψ (solid line), induced via the Yukawa interaction of ψ and $\tilde{\Phi}$ (dashed line) and the σ–$\tilde{\Phi}$ coupling given by (6.74), $\sim \zeta \lambda_\Phi m^2$.

where we here made $M_{\rm P}^{-1}$ explicit so that one can recognize dimensions correctly.

Suppose that the coefficient above were not $\cdots \zeta \lambda_\Phi m^2$, but instead were $\frac{1}{2}\zeta(m^2 + \delta m^2)$, with δm the self-mass of $\tilde{\Phi}$, which is divergent at the one-loop level. If this were the case, the source of σ would have been proportional to the total mass of the particle $\tilde{\Phi}$, implying that composition-independence holds at least for the one-particle state. Equation (6.74) shows that this is not in fact the case, however, resulting in the conclusion that the WEP is violated through the coupling (6.74).

According to the standard theory, the mass M of a fermion arises from the Yukawa interaction with Φ_*

$$-L_{\psi\Phi} = g_{\rm Y}\overline{\psi}\psi\Phi_*, \tag{6.75}$$

giving

$$M = g_{\rm Y}v. \tag{6.76}$$

Owing to the interaction (6.75), σ couples to ψ through the diagram in Fig. 6.2. The integral turns out to be finite. The result is represented by an effective Lagrangian given by

$$-L_{\psi\sigma 2} = g_\sigma \overline{\psi}\psi\sigma, \tag{6.77}$$

where the coupling constant is

$$g_\sigma = \frac{3}{2}\frac{M}{M_{\rm P}}\zeta\lambda_\Phi\frac{g_{\rm Y}^2}{4\pi}F_1\left(\frac{M^2}{m^2}\right), \tag{6.78}$$

with

$$F_1(x) = \frac{1}{6\pi^3}\frac{1}{x-1}\left(-\frac{5}{2} - \frac{1}{x-1} + \frac{x(2x-1)}{(x-1)^2}\ln x\right). \tag{6.79}$$

We also re-installed $M_{\rm P}^{-1}$ to demonstrate that g_σ is dimensionless.

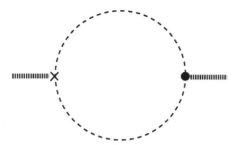

Fig. 6.3. A one-$\tilde{\Phi}$-loop diagram contributing to the self-energy of σ (thick, broken line). The cross is for the first term on the right-hand side of (6.56), $\zeta(D-4)m^2$, while the filled circle represents the coupling due to (6.74).

This g_σ is proportional to M, but depends also on g_Y, λ_Φ, and m/M, which represent the "internal structure" of ψ in a sense. Suppose that there are several different fermions. It is unlikely that the ratios of the coupling constants of these fields are precisely given by the ratios of the masses. This also indicates breakdown of the WEP.

As another consequence of the coupling of σ to ψ and Φ, the σ will easily acquire nonzero mass. Vector fields with gauge invariance, or fermions with chiral invariance, have built-in mechanisms that keep the fields massless. Unlike these fields, the scalar field σ has no "immunity" to protect it from acquiring a nonzero mass.

Consider the diagram in Fig. 6.3. At one of the vertices (a cross) we have $D-4$ coming from the first term on the right-hand side of (6.56), while we have another vertex (a filled circle) with the anomalous coupling given by (6.74); in this sense, the diagram comes originally from a two-loop diagram multiplied by $(D-4)^2$. We, however, simply follow the technique used for the previous calculations of anomalies and obtain

$$\mu_\sigma^2 = -i(2\pi)^{-D}4(2-d)\zeta\mu_\Phi^2\frac{\zeta\lambda_\Phi m^2}{\pi^2}I_c. \tag{6.80}$$

Using (6.71) we finally find, upon re-inserting M_P,

$$\mu_\sigma^2 = \frac{1}{4\pi^4}\zeta^2\lambda_\Phi\frac{m^4}{M_P^2}. \tag{6.81}$$

We notice, however, that there are some other diagrams that fail to reduce to those giving anomalies. We do have examples of one-loop diagrams in which both vertices remain nonvanishing at $D=4$. The integrals then exhibit quadratic divergences. Suppose, however, that approximate supersymmetry operates to suppress divergences. Then the integrals would be virtually cut off at M_{ssb}, the mass scale at which supersymmetry is broken, which is widely accepted to be somewhere around ~ 1 TeV. Consequently, σ would have a mass μ_σ given by

$$\mu_\sigma^2 \sim g_\sigma^2 M_{ssb}^2. \tag{6.82}$$

Once the dilaton ceases to be massless, the dilatation current is no longer conserved. Computation based on (6.54) gives in fact

$$\partial_\mu J^\mu = \zeta^{-1}\mu_\sigma^2 \sigma + v_\Phi m^2 \tilde{\Phi} + \cdots, \tag{6.83}$$

where $\cdots$ stands for other complicated terms, including terms like $\tilde{\Phi}^3 \sigma$ and $\tilde{\Phi}^4 \sigma$, in addition to those shown in (6.74).

Nonconservation of the current indicates that scale invariance is now broken *explicitly*. However, the violation will be "mild" if μ_σ is "small." The scalar field would then be called a *pseudo-NG boson*. It might be worth recalling that one of the best examples of pseudo-NG bosons is the massive pion, which motivated the whole idea of spontaneous breaking of symmetry in particle physics.

In our previous attempts to realize explicit breaking of scale invariance, we introduced certain dimensional parameters in a somewhat *ad hoc* manner [113]. We have now, however, a mechanism based on a quantum anomaly by which explicit breaking takes place almost imperceptibly. In relativistic quantum field theory, which includes many divergences, it is essentially unavoidable, for any sensible results, that one will have some dimensional quantities other than those present in the Lagrangian. We now realize that this provides a "natural" way to break scale invariance. We might even call it spontaneous breaking in a broader sense.

6.3.2 The coupling to matter of σ, scalar QED

So far we have demonstrated explicitly how the theory results in violation of the WEP. In (6.75) we also derived the coupling constant g_σ for the coupling of σ to a fermion field ψ. We want, however, to find the coupling constant for the coupling to nucleons for realistic applications. The foregoing analysis of the self-coupled scalar field Φ is obviously insufficient for this purpose. A better way seems to be to calculate instead the coupling of σ to the QCD system, and then compute the effective coupling constant for the coupling to nucleons as composite particles.

We have avoided this partly because we know no way to formulate the dilatation current for a spinor field, thus making it somewhat awkward to develop a consistent theoretical formulation. Another reason was that we must be familiar with the use of a "polyad" as an extension of the tetrad in four dimensions, though the extension can be made quite formally, if we confine ourselves to the task of evaluating loop integrals for $D \approx 4$. For these reasons, we develop the QED of a complex scalar field in this subsection, in order to access the realistic applications. This allows us to extend it to QCD rather easily. QED itself will also be a realistic model when we discuss the possible time-variation of the fine-structure constant in section 6.6.

The matter scalar field considered in section 6.2 is now extended to a complex scalar field, expressed by the same symbol Φ; the Hermitian conjugate is denoted by $\bar{\Phi}$. These are assumed to bear the charges e and $-e$, respectively. The electromagnetic field A_μ is also introduced. The matter Lagrangian in the J frame is thus given by

$$\mathcal{L}_{\text{matter}} = \sqrt{-g}\left(-g^{\mu\nu} \mathcal{D}^{(e)}_\mu \bar{\Phi}\, \mathcal{D}^{(e)}_\nu \Phi - \tfrac{1}{2}f^2\bar{\Phi}\Phi\phi^2 - \tfrac{1}{4}g^{\mu\nu}g^{\rho\sigma}F_{\mu\rho}F_{\nu\sigma}\right),$$

(6.84)

where

$$\mathcal{D}^{(e)}_\nu \Phi = (\partial_\nu - ieA_\nu)\Phi,$$ (6.85)

$$\mathcal{D}^{(e)}_\nu \bar{\Phi} = (\partial_\nu + ieA_\nu)\bar{\Phi},$$ (6.86)

$$F_{\mu\rho} = \partial_\mu A_\rho - \partial_\rho A_\mu.$$ (6.87)

Notice that we used the same appearance of the second term on the right-hand side of (6.1) as that in (4.129), which is slightly different from the corresponding term in (6.2), which is adapted more to the Higgs mechanism.

We have ignored the gauge-fixing term, though the calculation will be done in the Feynman gauge. As in subsection 6.3.1, we consider that spacetime is D-dimensional. We move to the E frame, by applying a conformal transformation,

$$g_{*\mu\nu} = \Omega^2 g_{\mu\nu}.$$ (6.88)

In the same way as with (4.80) for a neutral scalar field, we define the new complex field to make the kinetic term canonical:

$$\Phi_* = \Omega^{1-d}\Phi \quad \text{and} \quad \bar{\Phi}_* = \Omega^{1-d}\bar{\Phi}.$$ (6.89)

The scalar field Φ_* acquires a mass m_* according to the relation

$$m_*^2 = m^2 e^{(D-4)\zeta\sigma},$$ (6.90)

where

$$m^2 = \xi^{-1} f^2,$$ (6.91)

according to (4.131). Strictly speaking we have some more D-dependence in (6.90), as we can see in (6.34)–(6.36), though we simply used the values for $d = 2$, because they do not produce any crucial behavior at $d = 2$.

The transformation properties of A_μ need a little careful analysis. We find

$$\sqrt{-g}\,g^{\mu\nu}g^{\rho\sigma} = \sqrt{-g_*}\,\Omega^{4-D}g_*^{\mu\nu}g_*^{\rho\sigma}, \qquad (6.92)$$

which reminds us that the Maxwell term is conformally invariant in four dimensions. In D dimensions, however, the extra factor in (6.92) can be absorbed into the redefined field;

$$A_{*\mu} = \Omega^{2-d}A_\mu. \qquad (6.93)$$

The field strength is defined accordingly by

$$F_{*\mu\nu} = \Omega^{2-d}F_{\mu\nu} = [\partial_\mu A_{*\nu} + (d-2)(\partial_\mu \ln \Omega)A_{*\nu}] - (\mu \leftrightarrow \nu). \qquad (6.94)$$

We find the occurrence of the extra terms of derivatives of Ω and hence of σ, which, however, may be dropped in many practical applications.

We also find that the covariant derivatives in (6.85) and (6.86) can be redefined as

$$\mathcal{D}_{*\mu}^{(e)} = \partial_\mu \mp ie_*A_{*\mu} - (d-1)(\partial_\mu \ln \Omega), \qquad (6.95)$$

where we introduced e_* such that

$$eA_\mu = e_*A_{*\mu}. \qquad (6.96)$$

According to (6.93) we have

$$e_* = \Omega^{d-2}e = e\exp[(d-2)\zeta\sigma], \qquad (6.97)$$

which obviously plays the role of the electric charge in the E frame.

The matter Lagrangian (6.84) is now put into the form

$$\mathcal{L}_{\text{matter}} = \sqrt{-g_*}\left(-g_*^{\mu\nu}\,\mathcal{D}_{*\mu}^{(e)}\bar{\Phi}_*\,\mathcal{D}_{*\nu}^{(e)}\Phi_* - m_*^2\bar{\Phi}_*\Phi_* - \tfrac{1}{4}g_*^{\mu\nu}g_*^{\rho\sigma}F_{*\mu\rho}F_{*\nu\sigma}\right). \qquad (6.98)$$

We find that (6.98) contains terms that depend on σ through e_* and m_*^2. The electromagnetic interaction terms are immediately derived:

$$L_{\text{em1}} = -ie_*\bar{\Phi}_*\overset{\leftrightarrow}{\partial}_\mu\Phi_*A_*^\mu, \qquad (6.99)$$

$$L_{\text{em2}} = -e_*^2\bar{\Phi}_*\Phi_*A_{*\mu}A_*^\mu. \qquad (6.100)$$

Considering that σ is an external field that is almost constant, we obtain the source term S of σ, by linearizing (6.90) and (6.97), respectively:

$$m_*^2 \approx m^2[1 + 2(d-2)\zeta\sigma], \qquad (6.101)$$

$$e_* \approx e[1 + (d-2)\zeta\sigma]. \qquad (6.102)$$

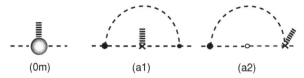

(0m)   (a1)   (a2)

Fig. 6.4. The gray disk in (0m) implies that radiative corrections are to be applied to the "skeleton" of the mass-squared insertion into Φ_* (dashed line), given by the right-hand side of (6.101). There are two one-loop diagrams, (a1) and (a2). The cross with the thick, broken line for σ is for the second term on the right-hand side of (6.101) and the second term on the right-hand side of (6.102), for (a1) and (a2), respectively. The open circle is for m^2 without the attached σ line, while the filled circle stands for the linear QED vertex given by (6.99). The dotted lines are for the electromagnetic field. To (a2) the right–left-symmetric diagram, with the filled circle and the cross interchanged, should be added.

The σ-dependence vanishes in the limit $D \to 4$, or $d \to 2$. This is true, however, *only classically*. In the following we show that the zero $(d-2)$ is going to be canceled out by the pole $(d-2)^{-1}$ which arises from loop integrals, thus yielding quantum anomalies.

First let us consider the diagrams represented by the one denoted by (0m) in Fig. 6.4, which originates from a tree diagram coming from m_*^2 in (6.101). At the one-loop level of QED, we have two diagrams corresponding to (a1) and (a2). In the first diagram, (a1), the σ-dependent part, $2(d-2)m^2\zeta\sigma$, is surrounded by a photon line, whereas the mass insertion in (a2) is simply a constant term m^2 while the charge in one of the electromagnetic vertices produces a term $2(d-2)e\zeta\sigma$. We add the "right–left-symmetric term" with the σ-dependent and -independent terms exchanged, which amounts simply to doubling one of the contributions.

We first discuss the contribution from diagram (a1). We must also consider the wave-function-renormalization terms, the diagrams corresponding to which will be denoted by (b1) and (c1) in Fig. 6.5. We do this on the basis of quantum field theory in tangential Minkowski space-time.

Diagram (a1) gives a source of σ:

$$S_{\mathrm{a1}} = -2i\zeta(2-d)m^2\frac{\alpha_*}{4\pi^3}I_2(-m^2),$$
(6.103)

where

$$I_2(p^2) = \int d^D k \, \frac{(k-2p)^2}{k^2[(k-p)^2+m^2]^2}.$$
(6.104)

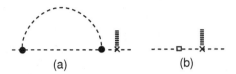

(a)   (b)

Fig. 6.5. Wave-function-renormalization diagrams, corresponding to the vertex diagram (a1) in Fig. 6.4. The small square in (b2) represents $-\delta m^2$. Taking the average with the right–left-symmetric term is understood. The diagrams corresponding to (a2) should be similarly added.

Fig. 6.6. The self-energy part of Φ_*.

We notice that the above integral is given by

$$I_2(p^2) = -\frac{\partial}{\partial m^2} I_1(p^2), \tag{6.105}$$

where

$$I_1(p^2) = \int d^D k \, \frac{(k - 2p)^2}{k^2[(k - p)^2 + m^2]}, \tag{6.106}$$

in terms of which the self-energy part of Φ_*, diagrammatically given by Fig. 6.6, is given by

$$\Sigma(p^2) = i\frac{\alpha_*}{4\pi^3} I_1(p^2). \tag{6.107}$$

On substituting this into (6.103), we obtain

$$S_{\mathrm{a}1} = -2\zeta(d - 2)\left[m^2 \frac{\partial}{\partial m^2}\Sigma(p^2)\right]_{p^2=-m^2}. \tag{6.108}$$

According to a custom in relativistic field theory, we expand $\Sigma(p^2)$ into a finite-order Taylor series:

$$\Sigma(p^2) = A(m^2) + \left(p^2 + m^2\right)B(m^2) + \left(p^2 + m^2\right)^2 \Sigma_f(p^2, m^2), \tag{6.109}$$

where

$$A(m^2) = \delta m^2 \tag{6.110}$$

is the self-mass squared.

Diagram (c1) in Fig. 6.5 comes from the mass counter-term $-\delta m^2$. After this subtraction from the self-energy diagram, $\Sigma(p^2)$ leaves $(p^2 + m^2)$ $B(m^2)$ (together with the finite term which is eventually ignored on the mass-shell). The factor $p^2 + m^2$ is canceled out by the "intermediate" propagator $(p^2 + m^2)^{-1}$, leaving $m^2 B$.

On the other hand, substituting (6.109) into (6.108) gives

$$S_{a1} = -2\zeta(2-d)\left(m^2 \frac{dA}{dm^2} + m^2 B + \cdots\right), \tag{6.111}$$

where $\cdots$ stands for terms that vanish on the mass-shell $p^2 + m^2 = 0$. The second term on the right-hand side of (6.111) is found to be canceled out by the wave-function-renormalization terms.

In this way we have

$$S_{m1} = S_{a1} + S_{b1} + S_{c1} = -2\zeta(2-d)m^2 \frac{dA}{dm^2}, \tag{6.112}$$

for (a1) and its associated wave-function renormalizations. We now evaluate $A(m^2)$, by putting $p^2 = -m^2$ in (6.107) and (6.106). We have

$$I_1(-m^2) = \int dx \int d^D \tilde{k} \, \frac{(k-2p)^2}{\mathcal{D}^2}, \tag{6.113}$$

where

$$\mathcal{D} = \tilde{k}^2 + \mathcal{M}^2, \tag{6.114}$$

with

$$\tilde{k} = k - xp,$$
$$\mathcal{M}^2 = x^2 m^2. \tag{6.115}$$

By performing symmetric integration, and using formulas in D-dimensional integrations, we obtain

$$A(m^2) = -\frac{\alpha}{4\pi} \int dx \left[\left(\mathcal{M}^2\right)^{d-1} B(d+1, 1-d)\right.$$

$$\left. - m^2(2-x)^2 \left(\mathcal{M}^2\right)^{d-2} B(d, 2-d)\right]$$

$$= -\frac{\alpha}{4\pi}\left(m^2\right)^{d-1} \int dx \left[x^{D-2} B(d+1, 1-d)\right.$$

$$\left. - (2-x)^2 x^{D-4} B(d, 2-d)\right]. \tag{6.116}$$

We find that the right-hand side of the second line contains $(m^2)^{d-1}$ as the sole dependence on m^2. From this it follows immediately that

$$m^2 \frac{dA}{dm^2} = (d-1)A. \tag{6.117}$$

The factor $d - 1$ can also be put equal to unity. Using this on the right-hand side of (6.112) hence gives

$$S_{m1} = -2\zeta(2 - d)A(m^2). \tag{6.118}$$

We move on to evaluate $(d - 2)A$. By using

$$(2 - d)B(1 + d, 1 - d) = (2 - d)\frac{\Gamma(1 + d)\Gamma(1 - d)}{\Gamma(2)}$$

$$= \frac{1}{1 - d}\frac{\Gamma(1 + d)}{\Gamma(2)}(2 - d)\Gamma(2 - d), \tag{6.119}$$

$$(2 - d)B(d, 2 - d) = \frac{\Gamma(d)}{\Gamma(2)}(2 - d)\Gamma(2 - d), \tag{6.120}$$

we find

$$(d - 2)A(m^2) \approx \frac{\alpha_*}{4\pi}m^2 \int dx \left[-2x^2 - (2 - x)^2\right](2 - d)\Gamma(2 - d)$$

$$= -\frac{3\alpha_*}{4\pi}m^2, \tag{6.121}$$

where we used

$$(2 - d)\Gamma(2 - d) \to 1 \tag{6.122}$$

in the limit $d \to 2$. The x integration has been performed also in this limit. Notice that (6.121) is *finite*. Since we are considering the linear source of σ, we may replace α_* in (6.121) simply by α. On substituting (6.121) into (6.118) we finally obtain

$$S_{m1} = -\frac{3\alpha}{2\pi}\zeta m^2. \tag{6.123}$$

We move on to compute diagram (a2) in Fig. 6.4 and its right–left-symmetric counterpart. We find that the contribution is given in terms of the same function I_2 in (6.104), and hence in terms of A, after taking the wave-function renormalization into account:

$$S_{m2} = 2\zeta(d - 2)m^2 \frac{dA}{dm^2}, \tag{6.124}$$

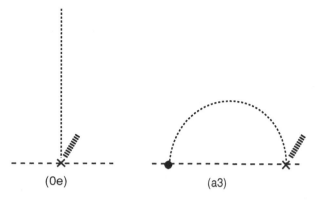

(0e)   (a3)

Fig. 6.7. A cross stands for the vertex to which a σ line is attached, while a filled circle is for the usual electromagnetic coupling.

where use of (6.108) and (6.109) has been made. Further using (6.117) with the setting $d = 2$ on the right-hand side, we obtain

$$S_{m2} = -2\zeta(2 - d)A, \qquad (6.125)$$

which is found to agree with (6.118):

$$S_{m1} = S_{m2} = -\frac{3\alpha}{2\pi}\zeta m^2. \qquad (6.126)$$

Another σ-coupling term arises directly from the self-energy part of Φ_*, by substituting $e_*(\sigma)$ into (6.99) and picking up terms linear with respect to σ, as shown in Fig. 6.7.

By using (6.102) we obtain

$$S_e = 2\zeta(d - 2)A = -2\zeta(2 - d)A, \qquad (6.127)$$

which again agrees with (6.118) and hence with (6.125). It then follows that all three kinds of diagrams contribute equally, giving

$$S = S_{m1} + S_{m2} + S_e = -3 \times \frac{3\alpha}{2\pi}\zeta m^2 = -\frac{9\alpha}{2\pi}\zeta m^2. \qquad (6.128)$$

Since these terms are the linear source of σ, we now have a term

$$\begin{aligned} -L_{\Phi\sigma} &= S\bar{\Phi}_*\Phi_*\sigma \\ &= -\frac{9\alpha}{2\pi}\zeta m^2\bar{\Phi}_*\Phi_*\sigma, \end{aligned} \qquad (6.129)$$

for the coupling of σ to the field Φ_*. The appearance of α on the right-hand side shows that the force mediated by σ is composition-dependent, and is against the WEP, for the same reason as has been elaborated in the preceding subsection.

This is a conclusion of this intermediary toy model. We add the remark that we have confined ourselves to the linear coupling term (6.99). We note, however, that the quadratic term in (6.100) results in a closed loop of a massless photon field, which vanishes for $D \neq$ integer.

6.4 Non-Newtonian gravity

We are now ready to translate the calculation to the coupling of σ to the QCD system. This can be done by applying the following simplified procedures.

- Replace Φ_* in the preceding subsection by a quark field q_*.

- Replace the photon by a set of color $SU(3)$ gauge fields $A_{*\mu}^a$ with $a = 1, 2, \ldots, N_c^2 - 1$, where $N_c = 3$.

- Replace every QED vertex by

$$ig\bar{q}_*\gamma^\mu T_a q_* A_{*\mu}^a, \tag{6.130}$$

  where g is a QCD coupling constant while T_a is an $N_c \times N_c$ matrix for a fundamental representation of $SU(N_c)$, satisfying $\mathrm{Tr}(T_a T_b) = \frac{1}{2}$.

- Make the contribution from the diagram (a1) be half, corresponding to the fact that the mass of a spinor field appears linearly rather than quadratically for a bosonic field.

- Multiply by the statistical factor,

$$\frac{N_c^2 - 1}{2N_c} = \frac{4}{3}, \tag{6.131}$$

  for $N_c = 3$.

- The mass squared m^2 of the scalar field is replaced by the linear mass m of a spinor mass for a quark.

We point out that QCD is a non-Abelian gauge theory, requiring more complications in estimating loop contributions [113]. We expect, however, that a simple extension of the Abelian calculation explained above will be sufficient to study results to a first approximation.

From (6.129), multiplied by $\frac{5}{6}$ as a consequence of the fourth remark above, and also by the factor (6.131), we obtain

$$-L_{\sigma q} = g_{\sigma q} m_q \bar{q}_* q_* \sigma, \qquad (6.132)$$

where the coupling constant $g_{\sigma q}$ is given by

$$g_{\sigma q} = \zeta \frac{5\alpha_s}{\pi} M_P^{-1} \approx 0.3 \zeta M_P^{-1}. \qquad (6.133)$$

Notice that we re-installed M_P^{-1}, making it explicit that $g_{\sigma q}$ has the dimension of inverse mass. Also α_s is a QCD analog of the fine-structure constant in QED, and is estimated to be ~ 0.2.

We want to see how this force mediated by the exchange of σ affects realistic physics. This will require us to have the force acting on nuclei, which make up most of the masses around us. It does not seem realistic to imagine that nuclei are made directly from quarks. We should obtain the coupling of σ to nucleons as important substructures. Viewing a nucleon as a system composed of quarks, we define the effective coupling constant $g_{\sigma N}$ by

$$g_{\sigma N} m_N = g_{\sigma q} \left\langle \sum_i m_{qi} \bar{q}_{*i} q_{*i} \right\rangle_N, \qquad (6.134)$$

where the summation extends to the colors and flavors of quarks [70]. We note that the above nucleon matrix element of the quark masses has been estimated to be ~ 60 MeV [115], amounting to only 6.4% of the total mass ~ 940 MeV. On combining this with (6.133), we find

$$g_{\sigma N} \sim 0.019 \zeta M_P^{-1}. \qquad (6.135)$$

This relatively small number is largely due to the smallness of the quark-mass component in the nucleon mass. One might suspect that a gluon component can make a considerable contribution. We point out, however, that we have included the effect of the coupling of σ to the gluon fields; (6.92) yielded a coupling to A_μ, but it has been converted into the coupling to the charge, whose effects have been included. More detailed estimation by going beyond the one-loop approximation on the basis of the renormalization-group approach is desired.

The smallness of the coupling constant estimated in (6.135) is crucially important for the following discussion. This favorable conclusion, however, might obviously be lost if we apply the same calculation directly to a nucleon, instead of the quarks. How can one justify the above procedure? In this connection we again remind readers that a nucleon is a composite object; momentum in a loop involving nucleons will be cut off at a certain

Quantum effects

small value without producing a divergence. After all, an anomaly is a "peaceful use" of an infinity such as that of nuclear energy [116].

Exchange of σ between two static nucleons then yields the Yukawa potential

$$V_\sigma(r) = -\frac{g_{\sigma N}^2}{4\pi} \left(\frac{m_N}{M_P}\right)^2 \frac{e^{-r/\lambda}}{r}, \tag{6.136}$$

where

$$\lambda = 1/\mu_\sigma, \tag{6.137}$$

is the force range.

According to (6.82) together with (6.132) and (6.133), we expect

$$\mu_\sigma^2 \sim m_q^2 M_{\text{ssb}}^2, \tag{6.138}$$

for the mass of σ given in terms of the quark mass and the mass scale for breaking of supersymmetry. By using the values of the masses for the present epoch, we find

$$\mu_\sigma \sim 0.84 \times 10^{-36} \approx 2.1 \times 10^{-9} \, \text{eV} \tag{6.139}$$

and the corresponding force-range

$$\lambda \sim 1.2 \times 10^{36} \approx 9.6 \times 10^3 \approx 100 \, \text{m}. \tag{6.140}$$

In view of the uncertainties in evaluating the self-energy of a particle, we should allow the estimate a few orders of magnitude latitude. All we should conclude from these simplified estimates is that the expected force-range is of a macroscopic size, and certainly much shorter than the distances in various aspects of the solar system.

More crucially, however, we may note that the field σ is subject to a cosmological potential $V(\sigma, \chi)$ as given by (5.58), giving a mass squared defined naturally by

$$\tilde\mu_\sigma^2 = \frac{\partial^2 V}{\partial \sigma^2}. \tag{6.141}$$

By simplifying the argument by choosing $\chi = 0$, and hence $V \sim e^{-4\zeta\sigma}$, we find

$$\tilde\mu_\sigma^2 \sim (4\zeta)^2 V. \tag{6.142}$$

Substituting this into the right-hand side of Einstein's equation (5.59), assuming that the potential term behaves in nearly the same way as other terms, and also assuming b' to be roughly of the order of unity (equivalently, assuming a power-law expansion of the universe), we finally obtain a crude estimate t_*^{-2} for the right-hand side of (6.142), assuming also that ζ remains of the order of unity. The result can be readily translated into

$$\lambda \sim t_*, \tag{6.143}$$

implying that the force-range obtained in this way is essentially the same as the size of the visible part of the whole universe, which is overwhelmingly different from the estimate given by (6.140) [94, 117, 118]. In the following we briefly argue why we favor (6.140) [119].

We first point out that the field σ can be broken up into two pieces:

$$\sigma = \sigma_{\rm b}(t_*) + \sigma_{\rm f}(x), \tag{6.144}$$

where the first term, $\sigma_{\rm b}(t_*)$, gives a cosmological background behavior, while the local fluctuation is described by the second term, $\sigma_{\rm f}(x)$, where x is for the coordinate on the local Lorentz frame. The force between two objects around us arises entirely from the local field $\sigma_{\rm f}$, which has quantum-theoretical interactions with other local fields, yielding a self-energy, as well, which has nothing to do with the cosmological environment.

According to an intuitive view, (6.143) simply tells us that the field is nearly massless insofar as the background part is concerned, and this fact does not prevent the field from acquiring the self-mass due to local interactions.

From a more rigorous point of view, we have an example of the presence of two different kinds of mass, as demonstrated by the sine-Gordon equation in two dimensions [100]. The quantized field has "mesonic excitations" at each of the sinusoidal potential minima, given by the second derivative of the local potential. Quite apart from them, there are classical soliton solutions that connect different potential minima. The mass of each such solution is in fact different from the mass of the "meson."

The analogy is far from complete in our present consideration. The cosmological solution does not seem to have a soliton behavior, though the equation is certainly nonlinear. What still interests us is that the soliton solution exhibits a behavior entirely different from the propagation of the mesonic excitation, which is a harmonic oscillator. In the same context we may allow a difference between μ_σ and $\tilde{\mu}_\sigma$.

In this way we use the above estimate (6.137) for the force-range, continuing the analysis of the potential (6.136). We use the relation

$$M_{\rm P}^{-2} = 8\pi G. \tag{6.145}$$

Substituting this into (6.136) allows us to write the potential in the form

$$V_\sigma(r) = -Gm_{\rm N}^2 \alpha_5 \frac{e^{-r/\lambda}}{r}, \tag{6.146}$$

where
$$\alpha_5 = 2g_{\sigma N}^2 \approx 0.72 \times 10^{-3}\zeta^2. \tag{6.147}$$

We are going to see that this value of the coefficient α_5 is rather close to the upper bounds obtained so far, though it appears somewhat too large by an order of magnitude or two, thus indicating the need for more accurate estimates.

In this connection we point out that we have by no means exhausted possible sources of composition-dependence of the σ coupling. No attention has been paid in this regard on the binding energies inside a nucleus, for example. This implies that we have no explicit way to show accurately how the relative strength of violation of the WEP changes from nucleus to nucleus. In order to obtain an approximate estimate of the scalar force, we only borrow the analysis based on the "fifth force" which is assumed to couple to the baryon number. Obviously, more extensive studies have to be applied to the QCD dynamics of "nuclear matter." With these reservations in mind, let us see how the extra potential (6.146) can be probed by phenomenological investigations.

With α_5 roughly of the order of unity, we should use the combined potential
$$V(r) = -G\frac{m_N^2}{r}\left(1 + \alpha_5 e^{-r/\lambda}\right), \tag{6.148}$$

to describe "gravitational phenomena."

Any macroscopic object may be regarded as made mostly of nucleons. Then the static potential for a force between two macroscopic objects with masses M_i and M_j separated by the distance r may be given by
$$V_{ij}(r) = -G\frac{M_i M_j}{r}\left(1 + \alpha_{5ij} e^{-r/\lambda}\right), \tag{6.149}$$

where we attach the suffices ij to α_5 as a reminder that the σ force is generally composition-dependent.

Suppose that the separation is much shorter than the force-range; $r \ll \lambda$. Then the Yukawa part behaves in the same way as the ordinary Newtonian potential, hence giving
$$V_{ij}(r) \approx -G_{ij}\frac{M_i M_j}{r}, \tag{6.150}$$

where
$$G_{ij} = G(1 + \alpha_{5ij}). \tag{6.151}$$

The force obeys simply the inverse-square law, but the coefficient G_{ij} is not only different from G, but also depends on the compositions of objects at both ends.

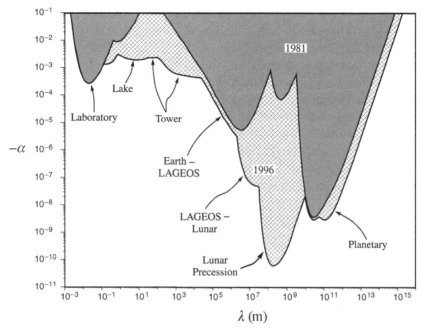

Fig. 6.8. Constraints on the coupling constant α (α_5 in our notation) as a function of the range λ from composition-independent experiments. The dark shaded area indicates the status as of 1981, and the lighter region the current limits, taken from Fig. 2.13 of [40]. © Springer-Verlag New York (1999).

On the other hand, the separation can be well beyond the force-range; $r \gg \lambda$. The second term is then dropped, leaving the Newtonian term alone:

$$V_{ij}(r) \approx -G\frac{M_i M_j}{r}. \tag{6.152}$$

The constant G provides the strength which controls the force in this long-distance limit.

In this way the potential (6.149) deviates from the purely inverse-square-law behavior of the gravitational force only for distances that are comparable to the force-range λ. This aspect of the potential has been the focus of experimental studies under the name of "composition-independent" experiments. Figure 6.8 [40] is a summary of the measurements performed so far. In this approach the composition-dependence of the coefficient α_5 can be mostly ignored.

We point out that this representation of a possible deviation from the purely Newtonian law in terms of the added Yukawa potential has been used recently as a convenient parameterization even for the millimeter range, for which other forms of deviation, such as the power-law behavior shown in (1.52), are expected [31].

Another type of approach for the non-Newtonian force is to probe for a possible violation of the WEP, in what are called "composition-dependent" experiments. How does the additional term, if there is one, affect free-fall phenomena on the Earth, for example? It seems to be a good idea first to obtain a rough order-of-magnitude estimate. If the potential is purely Newtonian, the distance r can be chosen to be that between the object and the center of the Earth. The acceleration from the Newtonian component is given by

$$g = G\frac{M_\oplus}{R_\oplus^2} = G\frac{4\pi}{3}\rho_{\mathrm{av}}R_\oplus, \tag{6.153}$$

where $M_\oplus$ and $R_\oplus$ are the Earth's mass and radius, respectively, while ρ_{av} represents the average density.

This simplification will be lost if the factor $e^{-r/\lambda}$ is present, though the contribution will obviously be much smaller. The calculation is simplified again if the height from the Earth's surface is much smaller than λ, which is also assumed smaller than $R_\oplus$. Since the finite-range force will come only from the matter distribution within a hemisphere of radius λ, we have the acceleration g_5 given roughly by

$$g_5 \approx \alpha_5 G\frac{1}{\lambda^2}\frac{2\pi}{3}\rho_{\mathrm{surf}}\lambda^3 = \alpha_5 G\frac{2\pi}{3}\rho_{\mathrm{surf}}\lambda, \tag{6.154}$$

where ρ_{surf} means the average density near the surface.

On comparing (6.153) and (6.154), we find

$$\frac{g_5}{g} \approx \frac{1}{2}\alpha_5\frac{\rho_{\mathrm{surf}}}{\rho_{\mathrm{av}}}\frac{\lambda}{R_\oplus}, \tag{6.155}$$

which simplifies to

$$\frac{g_5}{g} \sim \alpha_5\left(\frac{\lambda}{\mathrm{m}}\right) \times (4 \times 10^{-8}), \tag{6.156}$$

where we used the approximate estimate $\rho_{\mathrm{surf}}/\rho_{\mathrm{av}} \sim \frac{1}{2}$, while (λ/m) is for the value of λ measured in units of meters. In this way we learn that the effect of the finite-range component might not always be easily detectable even if $\alpha_5 \sim 1$, which is somewhat contrary to conventional wisdom.

So far, we have confined ourselves to the force between particles of the same kind. Modifications are also required if we consider forces between different kinds of particles, labeled by i and j, for example. The coupling constants for the coupling to σ may be written as $g_{\sigma i}$ and $g_{\sigma j}$, respectively. They are proportional commonly to $(m_i/M_P)\zeta$, as shown by (6.78), for example, which shows also that the remaining factors can differ from particle to particle, thus causing a composition-dependence of the force. The coefficient α_5 should then be replaced generally by α_{5ij} reflecting this dependence.

From the results of experimental studies carried out so far, however, the composition-dependence seems not to be very significant, if there is any. It then seems more convenient to separate the dependence into factors q_i with an overall factor α_5 representing a typical size. In this way we replace (6.149) by an extended form:

$$V_{ij}(r) = -\frac{Gm_i m_j}{r}\left(1 + \alpha_5 q_i q_j e^{-r/\lambda}\right). \qquad (6.157)$$

We consider q_i as a kind of "charge," having a magnitude roughly of the order of unity.

We may also consider that (6.147) or (6.149) applies to a force not only between elementary particles but also between macroscopic objects, with their masses M_i and charges q_i determined by properties of atoms and nuclei constituting the objects.

Since electrons are much lighter than nuclei, properties of objects will be dominated by those of nuclei. A nucleus made up of nucleons (protons and neutrons) has a mass given roughly by the sum of those of the nucleons. The nuclear force that binds nucleons together, however, gives a *binding energy*. We know that the binding energy per nucleon differs from nucleus to nucleus. The effect is rather significant, though the fractional differences are at the level of 10^{-3}. If the scalar field couples to nucleons and binding energy differently, the charge q_i can be different depending on the nucleus; $q_{Be} \neq q_{Cu}$, for example.

An explicit model of a *gravitational vector field* was proposed in order to illustrate the effect [120]. We briefly review this model, which is somewhat different from the scalar-field model we are considering, but has been discussed widely at least as an illuminating phenomenological approach.

Suppose that there is a vector field that couples to the baryon number. This implies that it couples to nucleons, equally to protons and neutrons, but not to any "field" responsible for the binding energy. Consider a nucleus i of atomic number A_i with mass M_i, hence yielding the binding energy B_i given by

$$M_i = mA_i - B_i, \qquad (6.158)$$

where m represents a nucleon mass, ignoring the difference in mass between a proton and a neutron for simplicity. The vector field couples to

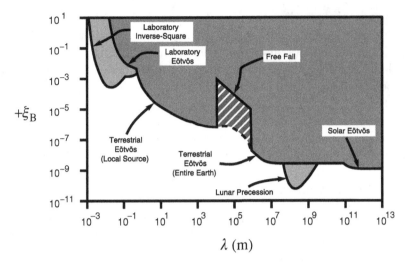

Fig. 6.9. Constraints on the coupling constant ξ_B as a function of the range λ from composition-dependent experiments, assuming a coupling to "baryon number," taken from Fig. 4.16 of [40]. © Springer-Verlag New York (1999).

this nucleus with the strength $\alpha_5 A_i$. We then obtain

$$q_i = \frac{mA_i}{M_i} \approx 1 + \frac{B_i}{mA_i}, \tag{6.159}$$

where we chose $M_i \approx mA_i$ in the last term, which, representing the binding energy per nucleon, differs from nucleus to nucleus.

Take examples: $B_{Be} = 6.5\,\text{MeV}$, $B_{Cu} = 8.5\,\text{MeV}$, and $B_{Au} = 7.9\,\text{MeV}$. Since $m \approx 940\,\text{MeV}$, we find $B_i/mA_i = 6.9 \times 10^{-3}, 9.04 \times 10^{-3}$, and 8.4×10^{-3} for Be, Cu, and Au, respectively. Differences among these numbers will be detected as an effect of violation of the WEP.

Consider that objects i and j made of nuclei i and j fall onto the Earth. Or they may fall toward another heavy object S. The difference in acceleration due to the extra term is given by

$$\frac{(\Delta g_5)_{ij,E}}{g} \approx \alpha_5 (\Delta q)_{ij} \left(\frac{\lambda}{\text{m}}\right) \times (4 \times 10^{-8}) \sim \alpha_5 \left(\frac{\lambda}{\text{m}}\right) \times 10^{-10}, \tag{6.160}$$

where we put $q_S \approx 1$, and chose $(\Delta q)_{ij} \sim 2 \times 10^{-3}$.

Figures 6.8 and 6.9 show the results on the strength α_5 from the composition-independent and -dependent experiments, as functions of the assumed λ. Notice that α and ξ_B (or ξ_I) can be identified roughly with our α_5.

For $\lambda \sim 1$ km, for example, Figs. 6.8 and 6.9 show the upper bounds $\sim 10^{-4}$–10^{-6}, though the results from composition-dependent experiments rely on the specific assumption of the baryon-number coupling. For the moment, we choose a typical size $|\alpha_5| \lesssim 10^{-5}$, for the sake of illustration. If we use this value in (6.160), we find

$$\frac{(\Delta g_5)_{ij,\mathrm{E}}}{g} \sim 10^{-12}, \tag{6.161}$$

giving an idea of how sensitive a measurement of the *difference* in acceleration we need in order to derive the result.

Many experiments searching for a difference between the accelerations of two falling objects have been done. They included high-precision measurements using torsion balances due originally to Eötvös *et al.* [120]. In spite of remarkable improvements in accuracies thanks to modern experimental technologies, however, no solid evidence for violation of the WEP has been obtained. Only upper bounds on the size of α_5 have been found, depending on the assumed value of λ [40, 111].

We point out that no precise estimate of q_i for the scalar force will be known until a more exact analysis in terms of quark dynamics becomes available.

6.5 A consequence of the suspected occurrence of the scalar field in the Maxwell term

In the calculation in the preceding sections, crucial ingredients were (6.101) and (6.102), for the mass and charge, respectively, which allowed us to calculate quantum effects unambiguously. The origin of these equations is traced back to the assumed scale invariance in the coupling to matter of ϕ for the mass combined with the conformal invariance of the Maxwell term for the charge, both in four dimensions. These invariances are based on the specific choice of how the scalar field enters the matter Lagrangian in the J frame.

In spite of the apparently successful results, however, they are not supported by such fundamental theories as string theory and KK theory, at least when such theories are interpreted naively. As was shown in Chapter 1, the factors $e^{-2\Phi}$ in the former, and $A^{n/2}\sqrt{\bar{g}}$ of "internal" space in the latter, contribute the same factor of ϕ commonly to the matter Lagrangian as the nonminimal coupling term in front of R. Rather disappointingly, we have no way out at present but to expect that compactification resulting in four-dimensional space-time involves some process that respects these invariances in a yet-to-be-elaborated version of the theories. One might also anticipate that our four-dimensional space-time was selected precisely for this reason. Without entering further into the question of how this is indeed to be implemented, we show in the following that the suspected occurrence of a function of the scalar field in front of the Maxwell term, or the kinetic term of the gauge fields, could be tested by composition-dependent experiments as considered in the preceding section.

Suppose that a function of ϕ is present in front of the last term on the right-hand side of (6.84). For a reasonable simplification, we choose it to be $\Omega^{2\iota}$:

$$\mathcal{L}_{\text{mx}} = -\tfrac{1}{4}\Omega^{2\iota}\sqrt{-g}\,g^{\mu\nu}g^{\rho\sigma}F_{\mu\rho}F_{\nu\sigma}, \tag{6.162}$$

where we use the four-dimensional relations

$$\xi^{1/2}\phi = \Omega = e^{\zeta\sigma}. \tag{6.163}$$

The choice $\iota = 1$ corresponds to what one expects naively from string theory as well as from KK theory, up to a multiplicative coefficient.

Equation (6.93) is then modified to

$$A_{*\mu} = \Omega^{2-d+\iota}A_\mu. \tag{6.164}$$

In accordance with this we have now

$$e_* = \Omega^{d-2-\iota}e \approx e[1 + (d-2-\iota)\zeta\sigma], \tag{6.165}$$

which replaces (6.97) and (6.102). Since ι is assumed nonvanishing, we may set $d = 2$, yielding

$$\alpha_* \approx \alpha(1 - 2\iota\zeta\sigma). \tag{6.166}$$

We warn the reader that ignoring quantum effects entirely is by no means justified. The effects are ambiguous because they depend on the possible high-energy cutoff, unlike in the anomaly-type calculation. We nevertheless will focus only on the classical effects, which can be evaluated with much less ambiguity.

Consider a Coulomb energy $\langle V_{\text{C}}\rangle_i$ of a nucleus i. Because it is proportional to α, which depends on σ in accordance with (6.166), we find that the Coulomb energy couples to σ as given by the interaction energy:

$$-L_{\text{C}\sigma} = -2\iota\langle V_{\text{C}}\rangle_i\zeta\sigma. \tag{6.167}$$

We further use the semi-empirical mass formula of Weizsäcker:

$$\langle V_{\text{C}}\rangle_i \approx m_{\text{C}}Z_i(Z_i - 1)A_i^{-1/3}, \quad \text{with} \quad m_{\text{C}} \approx 0.6\,\text{MeV}, \tag{6.168}$$

which may also be put into the form

$$-L_{C\sigma} \approx M_i\left(-2\iota\zeta\frac{m_C}{m_N}q_{Ci}\right)\sigma, \tag{6.169}$$

where

$$q_{Ci} = Z_i(Z_i - 1)A_i^{-4/3}, \tag{6.170}$$

by assuming $M_i \approx A_i m_N$ for the mass of the nucleus.

The σ-exchanged potential between two Coulomb energies is then expressed in the form of the second term on the right-hand side of (6.157):

$$V_{C5ij} = -\frac{GM_iM_j}{r}\alpha_{C5}q_{Ci}q_{Cj}e^{-r/\lambda}, \tag{6.171}$$

where

$$\alpha_{5C} = 2\left(2\iota\zeta\frac{m_C}{m_N}\right)^2 \approx 3.3 \times 10^{-6}(\iota\zeta)^2. \tag{6.172}$$

Compare this with the estimate $\alpha_5 \approx 0.72 \times 10^{-3}\zeta^2$, as given by (6.147), for a non-Newtonian force arising from the strong interaction. The right-hand side of (6.172) is about three orders of magnitude smaller if $\iota \sim 1$. The difference $(\Delta q_C)_{ij}$, however, can easily be of the order of unity, three orders of magnitude larger than what one expects from the binding energy of nuclei. As a consequence, the net effect for the differences in acceleration of two species of nuclei falling toward the Earth, $(\Delta g_{5C})_{ij,E}/g$, turns out to be nearly the same as that due to the strong interaction, (6.161).

The absence of any evidence for composition-dependence so far seems to indicate that the possibility of $\iota \sim 1$ is on the verge of being excluded. The present analysis contains much less ambiguity in determining q_{Ci} than does that explained in the preceding section. Further experimental searches from this point of view seem to be worth trying. We point out that the same kind of advantage over the strong interaction is not expected for composition-independent experiments.

6.6 Time-variability of the fine-structure constant

As another extension of the QED model discussed in the previous section, we now consider a possible time-variability of the fine-structure constant α, which is also a focus of the recent observational studies. This subject will be relevant for the following reasons.

We showed that the electric charge in the E frame couples to the scalar field σ in the way shown in (6.102) unless the classical contribution

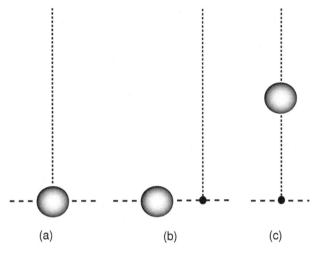

Fig. 6.10. Loop diagrams for quantum corrections to the electromagnetic coupling of Φ_*, represented by dashed lines. The dotted lines are for the electromagnetic field. Gray disks represent the vertex part and the self-energy part of Φ_*, and the photon self-energy part, for (a), (b), and (c), respectively. They might include the effect of the "strong" interaction, as well. Adding the right–left-symmetric diagram is understood for (b).

considered in the preceding section is present:

$$e_* \approx e[1 + \zeta(d - 2)\sigma]. \tag{6.173}$$

This by itself has no consequence classically for $d \rightarrow 2$, but has nonzero effects if it is combined with quantum corrections, as was discussed in connection with the possibility of violation of the WEP. Part of the calculation can be interpreted as a nontrivial dependence of α on σ. Since σ may vary with time, we should possibly expect a time-dependence of α, which can be compared with observations. We begin with an explicit computation of the σ-dependence of the electromagnetic coupling of the charged scalar field Φ_*.

6.6.1 Electromagnetic coupling of Φ_*

The electromagnetic coupling of Φ_* in the E frame is represented by the two terms in (6.99) and (6.100).

We first consider the one-loop diagrams Fig. 6.10 as radiative corrections to the linear term (6.99). For technical purposes, we may consider the e_*'s as if they were simply constant, disregarding their possible dependence on σ, for the moment. The diagrams fall into the three categories

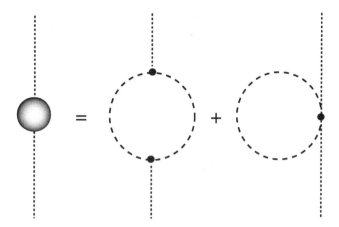

Fig. 6.11. The photon self-energy part to one-loop order.

illustrated in Fig. 6.10. The blobs in the vertex part (a) and the wave-function-renormalization terms (b) contain loops. We should also include the self-energy counter term $-\delta m_{*\Phi}^2$ in diagram (b). We find that all the terms of these two categories vanish or cancel each other out. This is due to the gauge invariance, assuring ultimately the universality of the electric charge. The conclusion is so general that it remains the same even if we include the contributions from other types of interaction, the strong interaction, for example.

We are thus left with the category (c), containing the photon self-energy part, giving the charge renormalization due to the vacuum polarization. In the present model the photon self-energy part, denoted by $\Pi_{\mu\nu}$, comes from two diagrams shown in Fig. 6.11, corresponding to (6.99) and (6.100), respectively.

The first diagram of Fig. 6.11 gives

$$\Pi_{\mu\nu}^{(1)}(k) = i(2\pi)^{-4}e_*^2 \int d^D p \, \frac{(2p+k)_\mu(2p+k)_\nu}{[(p+k)^2 + m_*^2](p^2 + m_*^2)}$$

$$= -\frac{\alpha_*}{4\pi} \int dx \left[\eta_{\mu\nu} \frac{2}{d} \left(\mathcal{M}^2\right)^{d-1} B(d+1, 1-d) \right.$$

$$\left. + (1-2x)^2 k_\mu k_\nu \left(\mathcal{M}^2\right)^{d-2} B(d, 2-d) \right], \quad (6.174)$$

where

$$\mathcal{M}^2 = m_*^2 + x(1-x)k^2. \quad (6.175)$$

We also assume that Φ_* has mass m_* either in the prototype BD model or in the scale-invariant model.

In the same way the second diagram of Fig. 6.11 gives

$$\Pi_{\mu\nu}^{(2)}(k) = -i(2\pi)^{-4}2e_*^2 \int d^D p \, \frac{1}{p^2 + m_*^2}$$

$$= \frac{\alpha_*}{2\pi}\eta_{\mu\nu}\left(m_*^2\right)^{d-1}B(d, 1-d). \tag{6.176}$$

We go to the limit $d \to 2$, finding

$$B(d+1, 1-d) = \frac{d+1}{1-d}\Gamma(2-d) \approx -2\Gamma(2-d), \tag{6.177}$$

$$B(d, 2-d) = \Gamma(d)\Gamma(2-d) \approx \Gamma(2-d), \tag{6.178}$$

$$B(d, 1-d) = \frac{\Gamma(d)\Gamma(1-d)}{\Gamma(1)} = \frac{\Gamma(d)}{1-d}\Gamma(2-d)$$

$$\approx -\Gamma(2-d), \tag{6.179}$$

which are substituted into (6.174) and (6.176). On adding the results together we finally obtain the gauge-invariant amplitude:

$$\Pi_{\mu\nu}(k) = \left(k_\mu k_\nu - \eta_{\mu\nu}k^2\right)C(k^2), \tag{6.180}$$

where

$$C(k^2) = -\frac{\alpha_*}{12\pi}\Gamma(2-d). \tag{6.181}$$

Given (6.180), diagram (c) of Fig. 6.10 contributes a term to the electromagnetic coupling of Φ_* given by

$$-e_*(2p+k)^\mu \frac{1}{k^2}\Pi_{\mu\nu}(k)a^\nu(k) = 2e_* Cp^\mu a_\mu(k), \tag{6.182}$$

where $a_\mu(k)$ is for the polarization vector of A_μ. This equation should be interpreted as an extra charge δe_* of the external line of Φ_* generated by the vacuum polarization due to the Φ_* loop. Since δe_* is supposed to contribute the amplitude

$$2\,\delta e_*\, p^\mu a_\mu(k), \tag{6.183}$$

we come to the relation

$$\delta e_* = e_* C = -\frac{\alpha_*}{12\pi}e_*\Gamma(2-d). \tag{6.184}$$

We now take the σ-dependence of e_* into account. We re-express (6.184) as

$$\delta e_*(\sigma) = -\frac{e_*^3(\sigma)}{48\pi^2}\Gamma(2-d), \tag{6.185}$$

and substitute (6.173) into each of the three e_*'s on the right-hand side.

Consider two different times t_1 and t_2 when the time-dependent σ is σ_1 and σ_2, respectively. Equation (6.173) gives the *classical* difference $\tilde{\Delta}e_*$ of the charge corresponding to the difference of σ:

$$\tilde{\Delta}e_* \approx e_*(d-2)\zeta\,\Delta\sigma, \tag{6.186}$$

where e on the right-hand side of the original (6.173) has been replaced approximately by e_*, and

$$\Delta\sigma \equiv \sigma_2 - \sigma_1. \tag{6.187}$$

We evaluate the right-hand side of (6.185) for σ_1 and σ_2, obtaining the difference of δe_* for σ_1 and σ_2 on the left-hand side. This difference in the charge due to the *quantum-theoretical* vacuum polarization is naturally interpreted as the difference in the whole charge itself:

$$\delta e_*(\sigma_2) - \delta e_*(\sigma_1) = e_*(\sigma_2) - e_*(\sigma_1) \equiv \Delta e_*. \tag{6.188}$$

For the classical difference of the right-hand side of (6.185) we use

$$\tilde{\Delta}\left(e_*^3\right) \approx 3e_*^2\,\tilde{\Delta}e_*. \tag{6.189}$$

Using (6.188) and (6.189) on both sides of (6.185), we derive

$$\Delta e_* = -e_*^2\frac{\tilde{\Delta}e_*}{16\pi^2}\Gamma(2-d)$$

$$\approx -\frac{\alpha_*}{4\pi}e_*(d-2)\zeta\,\Delta\sigma\,\Gamma(2-d). \tag{6.190}$$

Again using

$$(2-d)\Gamma(2-d) \to 1, \quad \text{as} \quad d \to 2, \tag{6.191}$$

we finally obtain

$$\frac{\Delta e_*}{e_*} = \frac{\alpha_*}{4\pi}\zeta\,\Delta\sigma, \tag{6.192}$$

or

$$\frac{\Delta\alpha_*}{\alpha_*} = \frac{\alpha_*}{2\pi}\zeta\,\Delta\sigma. \tag{6.193}$$

We have derived (6.181) and hence (6.193) according to a "toy model" of a charged scalar field Φ_*. Are these the same for any other types of fields? Do the contributions from other fundamental fields add up or cancel out?

Our calculation on a spinor field shows that the result is precisely the same. Adding up three flavors and three colors of quarks will multiply the right-hand side of (6.193) by a factor of two. Charged leptons will also contribute, resulting in the total multiplicative factor being 5.

We note that the present result shares the same nature as what is called the "trace anomaly," which was obtained originally from the non-Abelian gauge fields. The result for the Abelian gauge field is, however, not exactly the same as the trace anomaly. The sign on the right-hand side of (6.193) reflects, nevertheless, the fact that the photon is asymptotically unfree, providing a reason for why a spinor field contributes the same sign as the scalar field. We may also argue that the $W^\pm$ do not contribute to the loop because of their asymptotically free nature, though an opposite sign might arise because of the "wrong" relation between spin and statistics for the Faddeev–Popov ghost. We probably need a more careful re-examination of the issue also because the photon is not purely Abelian. We still appreciate the success of QED, which has been supported by such phenomena as the Lamb shift and the anomalous magnetic moment of the electron. In view of these uncertainties, we tentatively assume that (6.193) is multiplied by a parameter $\mathcal{Z}$:

$$\frac{\Delta\alpha_*}{\alpha_*} = \mathcal{Z}\frac{\alpha_*}{2\pi}\zeta\,\Delta\sigma \approx 1.2\mathcal{Z} \times 10^{-3}\,\Delta\sigma. \qquad (6.194)$$

A naively expected value is $\mathcal{Z} = 5$, as stated above. At this moment, we expect that $\mathcal{Z}$ is not much different from this value. We also chose ζ to be of the order of unity.

6.6.2 The behavior of the scalar field

What is the behavior of the scalar field σ which we should use to predict the time-variability of α? Obviously $\sigma(t)$ should be such a solution of the cosmological equation that fits other observational facts, particularly the accelerating nature of the universe.

We have designed models of scalar–tensor theory such that we can understand a small but nonzero cosmological constant, described effectively by $\Omega_\Lambda \sim 0.7$. We showed that a plateau behavior of the scalar-field energy may mimic the desired cosmological constant. This implies that the scalar field itself is nearly constant around the present epoch. If σ stays completely constant in time, then no time-variation of α is expected from (6.193) or (6.194). This seems to be too naive a view, however.

In fact a typical solution illustrated in Fig. 5.8 of Chapter 5 shows that $\sigma(t)$ is quite flat around the present epoch. A magnified plot, Fig. 5.10, reveals, however, a small oscillatory behavior. How seriously should this be accepted? Does a slight change of parameters erase it?

From the analysis we performed in connection with Fig. 5.14, we find that this behavior represents the fact that the scalar field is falling toward a minimum of the temporary potential. The amplitude of this oscillation depends on how close σ during the hesitation period had been to the potential minima. In this sense, an oscillatory behavior is generic to the present dynamical model of the scalar fields.

Let us accept the solution shown in Fig. 5.10, for the moment, and try to compare it with the reported result due to [18]. Their oldest quasi-stellar object is supposed to record the frequencies at the look-back time ~ 0.9 (of the age of the universe), which corresponds to $\hat{\tau} = \log t_* = \hat{\tau}_0 - 1$, where $\hat{\tau}_0 \approx 60$ for the present epoch. The time-dependence of α seems to be nonmonotonic, or even oscillating. This view might be reinforced if we add the results from the Oklo phenomenon [122–124], which are about two orders smaller, though the look-back time is also smaller (~ 0.16).

According to Fig. 5.10, $\Delta\sigma$, the difference in the value of σ with reference to the present value, seems to reach a maximum of 0.04. Suppose that this value corresponds to $\Delta\alpha/\alpha \approx 1.2 \times 10^{-5}$ [18], disregarding the sign for the moment. Substituting these values into (6.194) yields

$$\mathcal{Z} \sim \tfrac{1}{4}, \tag{6.195}$$

which seems to be acceptable in view of the crude approximation we have made, though the result in (6.195) is an order of magnitude smaller than the value $\mathcal{Z} = 5$ expected from a naive estimate coming from the fundamental fermions.

Before a complete analysis, which has yet to be done, it is worth pointing out that our discussion has made it possible to understand for the first time why the rate of change, or the upper bound, could be several orders of magnitude smaller than a "natural" guess of $\sim 10^{-10}\,\text{years}^{-1}$.

Looking at Fig. 5.14 more closely shows, as we mentioned, that we may choose the initial value σ_1 slightly differently, obtaining less oscillation but still keeping the agreement with the cosmological observation relatively unaffected. At the present preliminary stage of the analysis, it appears that measurements of $\Delta\alpha/\alpha$ impose severer restrictions than does the observation of the accelerating universe.

From the way the oscillation of σ dies out toward the present epoch, we may also expect that the smaller value of $\Delta\alpha$, if any, suggested by results from the Oklo phenomenon can be understood. The latter result obtained only for a relatively short duration may also correspond nearly to a zero of the oscillation.

We re-emphasize that we have started by assuming a purely constant α in the J frame, a theoretical conformal frame according to our previous conjecture. Admitting that this assumption itself is subject to further justification, we point out that the very fact that we can play this intriguing game demonstrates an obvious advantage of working with such a constrained framework as the scalar–tensor theory.

In the preceding analysis we started by choosing the E frame as a physical conformal frame, in which G as well as the mass of the field Φ_* is time-independent. The latter statement is true, however, only classically. At the level of quantum theory at which we discussed time-variation of the charge e, we should include the same for the mass, given by

$$m_*^2 = m^2(1 - 2\zeta'\sigma), \tag{6.196}$$

where

$$\frac{\zeta'}{\zeta} = \frac{9\alpha}{4\pi} \approx 5.2 \times 10^{-3}, \tag{6.197}$$

which is rather small.

These relations can be obtained by adding (6.129) to (6.91). Mass m_* changes as σ does. We have ignored this effect because it will have only a secondary importance practically when we focus on the change of α. In principle, however, we must make another conformal transformation from the point of view that we use clocks and meter-sticks on the basis of particle masses. After we move to the new conformal frame, which we call conveniently the J′ frame, we are going to have time-dependent G'. Details of the analysis will be found in Appendix O.

By exploiting the fact that both G' and α would change by a common origin $\delta\sigma$, we derive the relation

$$\frac{\Delta G'}{G'} \approx -9\mathcal{Z}^{-1}\frac{\Delta\alpha}{\alpha}, \tag{6.198}$$

which is somewhat similar to, but not exactly the same as, the relations in the literature [125]. We note that the linear relation above is only approximate because the value of $\Delta\sigma$ is also affected by the conformal transformation, which we have ignored because the effect is small.

In the context of the present analysis, we expect at most $\dot{G}'/G' \sim 10^{-14}\,\text{years}^{-1}$, somewhat below the available upper bounds [49, 50], and probably a goal of future measurements.

Appendix A
The scalar field from Kaluza–Klein theory

Consider the Einstein–Hilbert term in D-dimensional space-time:

$$\mathcal{L} = \tfrac{1}{2}\mathcal{C}\sqrt{-^{(D)}g}\,^{(D)}R, \qquad (A.1)$$

where $\mathcal{C}$ is a constant having mass dimension $D-2$ in the present system of units, represented by $1/(8\pi G_D)$ with (1.42) in subsection 2.3. Imagine also that the $n\,(=D-4)$-dimensional part is compactified to a small space, leaving the four-dimensional portion as infinitely extended space-time.

We start by assuming space-time of D dimensions. We consider the metric given by $g_{\bar{\mu}\bar{\nu}}$, where $\bar{\mu}$ and $\bar{\nu}$ run through the extended range $0, 1, \ldots, D-1$. The Christoffel symbol is defined formally in the same way as usual;

$$\Gamma^{\bar{\rho}}_{\bar{\mu}\bar{\nu}} = \tfrac{1}{2}g^{\bar{\rho}\bar{\sigma}}(\partial_{\bar{\mu}}g_{\bar{\sigma}\bar{\nu}} + \partial_{\bar{\nu}}g_{\bar{\sigma}\bar{\mu}} - \partial_{\bar{\sigma}}g_{\bar{\mu}\bar{\nu}}). \qquad (A.2)$$

Many of the formulas in Appendix C apply with indices $\mu, \nu, \ldots$ replaced by $\bar{\mu}, \bar{\nu}, \ldots$. According to our notation system, we should have written $^{(D)}g_{\bar{\mu}\bar{\nu}}$ and $^{(D)}\Gamma^{\bar{\rho}}_{\bar{\mu}\bar{\nu}}$, etc., but we sometimes suppress the superscript (D) to keep the expressions shorter, as long as correct understanding can be taken for granted.

To implement partial compactification, we make the *Ansatz* that $g_{\bar{\mu}\bar{\nu}}$ can be put into a block form;

$$g_{\bar{\mu}\bar{\nu}} = \begin{pmatrix} g_{\mu\nu} & 0 \\ 0 & g_{\alpha\beta} \end{pmatrix}, \qquad (A.3)$$

where $\mu, \nu, \ldots$ are four-dimensional while $\alpha, \beta, \ldots$ run from 1 to n for the coordinates θ_α. Off-block components would provide 4-vectors in four-dimensional space-time, but are ignored for the moment in order to focus on the scalar field. We also write

$$g_{\alpha\beta} = \mathcal{A}(x)\tilde{g}_{\alpha\beta}(\theta) \quad \text{and} \quad g^{\alpha\beta} = \mathcal{A}(x)^{-1}\tilde{g}^{\alpha\beta}(\theta), \tag{A.4}$$

where

$$\mathcal{A} = A^2, \tag{A.5}$$

where $A(x)$, assumed to be a function of the four-dimensional coordinate x, represents the size of n-dimensional "internal" space described by a purely geometrical metric $\tilde{g}_{\alpha\beta}$. Note that A is a length, having the mass dimension of –1. This implies that $\tilde{g}_{\alpha\beta}$ and the coordinates are dimensionless.

The assumed lack of dependence of A on θ's corresponds to limiting ourselves to zero modes, ignoring higher modes with masses of the order of M_P. It follows that

$$\sqrt{-^{(D)}g} = \sqrt{-g}A^n\sqrt{\tilde{g}}. \tag{A.6}$$

Accepting (A.3), we compute connections, finding nonzero components given by

$$\left.\begin{aligned}
^{(D)}\Gamma^\lambda_{\mu\nu} &= \Gamma^\lambda_{\mu\nu}, \\
\Gamma^\lambda_{\alpha\beta} &= -\tfrac{1}{2}\tilde{g}_{\alpha\beta}g^{\lambda\rho}\,\partial_\rho\mathcal{A}, \\
\Gamma^\alpha_{\mu\beta} &= \tfrac{1}{2}\delta^\alpha_\beta\mathcal{A}^{-1}\,\partial_\mu\mathcal{A}, \\
\Gamma^\alpha_{\beta\gamma} &= \tilde{\Gamma}^\alpha_{\beta\gamma},
\end{aligned}\right\} \tag{A.7}$$

where a tilde means a quantity computed in terms of $\tilde{g}_{\alpha\beta}$ alone. Corresponding to (C.5) and (C.6) we derive

$$^{(D)}\Gamma_\mu = \Gamma_\mu + \tfrac{1}{2}n\mathcal{A}^{-1}\,\partial_\mu\mathcal{A}, \tag{A.8}$$

$$^{(D)}\Gamma_\alpha = \tilde{\Gamma}_\alpha, \tag{A.9}$$

and

$$^{(D)}C^\lambda = C^\lambda - \tfrac{1}{2}n\mathcal{A}^{-1}g^{\lambda\mu}\,\partial_\mu\mathcal{A}, \tag{A.10}$$

$$^{(D)}C^\alpha = A^{-1}\tilde{C}^\alpha, \tag{A.11}$$

where quantities without a superscript are those computed purely four-dimensionally, unless obviously understood otherwise.

A D-dimensional extension of (C.4) then reads

$$^{(D)}R_2 = R_2 + \mathcal{A}^{-1}\tilde{R}_2 + \tfrac{1}{2}n\mathcal{A}^{-1}G^\mu\,\partial_\mu\mathcal{A} - \tfrac{1}{4}n(n-1)\mathcal{A}^{-2}g^{\mu\nu}\,\partial_\mu\mathcal{A}\,\partial_\nu\mathcal{A}, \tag{A.12}$$

where G^μ is defined by (C.9), with C^λ and Γ_μ defined by purely four-dimensional quantities.

We obtain the effective four-dimensional Lagrangian by integrating the D-dimensional Lagrangian over n-dimensional space:

$$\mathcal{L}_4 = \tilde{\mathcal{V}}_n^{-1} \int \mathcal{L} \, d^n\theta \tag{A.13}$$

$$= \mathcal{C}\sqrt{-g}\,a^n \tilde{\mathcal{V}}_n^{-1} \int \tfrac{1}{2}\sqrt{\tilde{g}}\,{}^{(D)}R \, d^n\theta, \tag{A.14}$$

where we divided conveniently by the "volume"

$$\tilde{\mathcal{V}}_n = \int \sqrt{\tilde{g}} \, d^n\theta \tag{A.15}$$

of the compactified space apart from A^n. Notice that the true volume V_n is given by

$$V_n = A^n \tilde{\mathcal{V}}_n. \tag{A.16}$$

We substitute (A.12) together with (A.6) into (A.14) with R replaced by $-R_2$. The third term on the right-hand side of (A.12) is found, after integration by parts, to contribute, apart from a 4-divergence,

$$-\tfrac{1}{2}\sqrt{-g}\,(\partial_\mu A^n)\,G^\mu = \tfrac{1}{2}A^n\,\partial_\mu(\sqrt{-g}G^\mu), \tag{A.17}$$

which is combined with the contribution from the first term on the right-hand side of (A.12), giving

$$\tfrac{1}{2}\sqrt{-g}\,A^n R, \tag{A.18}$$

where we used

$$\sqrt{-g}R = \partial_\mu(\sqrt{-g}G^\mu) - \sqrt{-g}R_2, \tag{A.19}$$

as obtained by combining (the purely four-dimensional versions of) (A.2) and (A.8).

The fourth term on the right-hand side of (A.12) appears to provide a term reminiscent of a kinetic term of the field $\mathcal{A}$ or A. We try the identification

$$-\tfrac{1}{2}\mathcal{C}\sqrt{-g}\,A^n\left[-\tfrac{1}{4}n(n-1)\mathcal{A}^{-2}g^{\mu\nu}\,\partial_\mu\mathcal{A}\,\partial_\nu\mathcal{A}\right] = \tfrac{1}{2}\sqrt{-g}g^{\mu\nu}\,\partial_\mu\phi\,\partial_\mu\phi. \tag{A.20}$$

For $n > 1$, (A.20) suggests

$$\tfrac{1}{2}\mathcal{C}^{1/2}A^{n/2}\sqrt{n(n-1)}\mathcal{A}^{-1}\,d\mathcal{A} = \mathcal{C}^{1/2}\sqrt{n(n-1)}A^{n/2-1}\,dA = d\phi, \tag{A.21}$$

which solves to give

$$2C^{1/2}\sqrt{\frac{n-1}{n}}A^{n/2} = \phi, \quad \text{or} \quad CA^n = \frac{1}{4}\frac{n}{n-1}\phi^2. \qquad (A.22)$$

From the mass dimension of C mentioned at the beginning of this appendix, one easily finds that ϕ defined in this way turns out to have the correct mass dimension of 1.

It should be noticed, however, that the sign on the right-hand side of (A.20) indicates that ϕ is a *ghost* field having a negative energy. The origin of this "wrong" sign can be seen already in (A.12). We remind the reader, however, that we show in Chapters 1 and 2 that this does not necessarily imply an immediate physical inconsistency thanks to mixing with the spinless component of the metric field. For $n = 1$, namely $D = 5$, there is no kinetic term of the scalar field, but we still have the dynamical degree of freedom of the field A, which is also explained in Chapters 1 and 2.

Expressing the scalar field in terms of ϕ and summing up all the terms in (A.12), we finally obtain the Lagrangian density for $n > 1$;

$$L_4 = \frac{1}{8}\frac{n}{n-1}\phi^2 R + \frac{1}{2}g^{\mu\nu}\partial_\mu\phi\,\partial_\nu\phi + \frac{1}{2}\left(\frac{1}{4}\frac{n}{n-1}\phi^2\right)^{1-2/n}\tilde{R}. \qquad (A.23)$$

The first two terms appear the same as the first two terms of (1.12) for the prototype BD model with the choices

$$\xi = \frac{1}{4}\frac{n}{n-1}, \quad \text{with} \quad \epsilon = -1, \quad \text{for} \quad n > 1. \qquad (A.24)$$

As it turns out, the condition (2.62) or (3.30), $\zeta^{-2} = 6 + \epsilon\xi^{-1} > 0$, for positive energy is always satisfied.

For the exceptional choice of $n = 1$ we simply choose

$$A = \phi^2, \qquad (A.25)$$

which is consistent with the condition $A > 0$, hence yielding

$$L = \tfrac{1}{2}\phi^2 R, \qquad (A.26)$$

where we dropped the term of $\tilde{R}$ which vanishes for $n = 1$. One might multiply the right-hand sides of (A.25) and (A.26) by an *arbitrary* coefficient ξ, obviously without affecting any further result. For example, (1.13) gives $\zeta^{-2} = 6$ for $\epsilon = 0$ independently of the value of ξ.

One may wonder why the KK theory entails the prototype BD model. In this connection, we notice that the Christoffel symbol has length dimension -1 coming from the derivatives of the coordinates having length dimension. The length dimension comes from A, on the other hand, for dimensionless coordinates θ_α. As a result, each term on the right-hand side of (A.12) has length dimension –2 coming either from ∂_μ or from A^{-1}, without the intervention of any dimensional constants. According to this, the first two terms on the right-hand side of (A.23), both of which contain two derivatives, should share the same term ϕ^2 as that which supplies length dimension –2.

Also worth noticing is that ξ given by (A.24) is larger than $\frac{1}{4}$ for nonzero n, which is obviously different from the very small value expected from the observations.

Appendix B

The curvature scalar from the assumed two-sheeted space-time

In Appendix A, we analyzed the five-dimensional KK theory with the local radius $A(x)$ in terms of the fundamental Einstein–Hilbert term with the second derivatives removed. In order, however, to make a comparison with the original form used by Saito et $al.$ [38], we re-analyzed the previous calculation but now with the second derivatives included in the curvature scalar. We start with

$$^{(5)}R^{\bar{\mu}}{}_{\bar{\nu}\bar{\rho}\bar{\sigma}} = \partial_{\bar{\rho}}\Gamma^{\bar{\mu}}{}_{\bar{\nu}\bar{\sigma}} + \Gamma^{\bar{\mu}}{}_{\bar{\lambda}\bar{\rho}}\Gamma^{\bar{\lambda}}{}_{\bar{\nu}\bar{\sigma}} - \text{(terms with } \bar{\rho} \leftrightarrow \bar{\sigma}). \qquad (\text{B.1})$$

According to (A.7), we already have

$$\left.\begin{array}{l} ^{(5)}\Gamma^{\lambda}{}_{\mu\nu} = \Gamma^{\lambda}{}_{\mu\nu}, \\[2mm] ^{(5)}\Gamma^{\lambda}{}_{44} = -\frac{1}{2}g^{\lambda\rho}\partial_{\rho}\mathcal{A}, \\[2mm] ^{(5)}\Gamma^{4}{}_{4\mu} = \frac{1}{2}\mathcal{A}^{-1}\partial_{\mu}\mathcal{A}, \end{array}\right\} \qquad (\text{B.2})$$

for five dimensions.

The nonzero components of (B.1) turn out to be

$$^{(5)}R^{\mu}{}_{\nu,\rho\sigma} = R^{\mu}{}_{\nu,\rho\sigma}, \qquad (\text{B.3})$$

and

$$\begin{aligned} ^{(5)}R^{\mu}{}_{4,\mu 4} &= \partial_{\mu}\Gamma^{\mu}{}_{44} - \partial_{4}\Gamma^{\mu}{}_{4\mu} + \Gamma^{\mu}{}_{\lambda\mu}\Gamma^{\lambda}{}_{44} - \Gamma^{\mu}{}_{44}\Gamma^{4}{}_{4\mu} \\[2mm] &= -\partial_{4}\Gamma^{\mu}{}_{4\mu} - \frac{1}{2}\nabla_{\mu}(\partial^{\mu}\mathcal{A}) + \frac{1}{4}\mathcal{A}^{-1}(\partial\mathcal{A})^{2}, \end{aligned} \qquad (\text{B.4})$$

$$\begin{aligned} ^{(5)}R^{4}{}_{\mu,4\nu} &= \partial_{4}\Gamma^{4}{}_{\mu\nu} - \partial_{\nu}\Gamma^{4}{}_{4\mu} + \Gamma^{4}{}_{\lambda 4}\Gamma^{\lambda}{}_{\mu\nu} - \Gamma^{4}{}_{4\nu}\Gamma^{4}{}_{\mu 4} \\[2mm] &= \partial_{4}\Gamma^{4}{}_{\mu\nu} - \frac{1}{2}\nabla_{\nu}\left(\mathcal{A}^{-1}\partial_{\mu}\mathcal{A}\right) - \frac{1}{4}\mathcal{A}^{-2}(\partial_{\nu}\mathcal{A})(\partial_{\mu}\mathcal{A}). \end{aligned} \qquad (\text{B.5})$$

In the KK theory in Appendix A, the first terms containing ∂_4 in the last lines of (B.4) and (B.5) vanish because the metric in five dimensions is assumed to be made independent of the fifth coordinate by ignoring higher-mass states. This is the place, however, where a difference enters the present theory of $M_4 \times Z_2$. We try first to estimate $-\partial_4 \Gamma^\mu_{\;4\mu}$ in the first term of the last line of the right-hand side of (B.4).

As in parallel transport in the direction of continuous spaces, we consider the change in the basis vectors,

$$\vec{e}_{\bar{\mu}}(x, g + r) = \vec{e}_{\bar{\mu}}(x, g) + \vec{e}_{\bar{\rho}}(x, g)\Gamma^{\bar{\rho}}_{\;\bar{\mu}4}\,\Delta w. \tag{B.6}$$

It then follows that

$$\Delta g_{\bar{\mu}\bar{\nu}} = (\Gamma_{\bar{\mu},\bar{\nu}4} + \Gamma_{\bar{\nu},\bar{\mu}4})\,\Delta w + \Gamma_{\bar{\rho},\bar{\nu}4}\Gamma^{\bar{\rho}}_{\;\bar{\mu}4}(\Delta w)^2, \tag{B.7}$$

and hence

$$\partial_4 g_{\bar{\mu}\bar{\nu}} = \Gamma_{\bar{\mu},\bar{\nu}4} + \Gamma_{\bar{\nu},\bar{\mu}4} + \Gamma_{\bar{\rho},\bar{\nu}4}\Gamma^{\bar{\rho}}_{\;\bar{\mu}4}\,\Delta w. \tag{B.8}$$

The coordinate condition can be derived only in the limit $\Delta w \to 0$. For this reason we might drop the last term on the right-hand side of (B.8). We allow this term to remain, however, and differentiate both sides of (B.8) with respect to w.

On the left-hand side we obtain a second derivative with respect to w, yielding zero, because the components of the metric are linear functions of Δw at most. An explicit illustration can be found in the relation (1.61):

$$\partial_4 \Delta w(x^4) = -2. \tag{B.9}$$

In this context we obtain

$$0 = \partial_4(\Gamma_{\bar{\mu},\bar{\nu}4} + \Gamma_{\bar{\nu},\bar{\mu}4}) - 2\Gamma_{\bar{\rho},\bar{\nu}4}\Gamma^{\bar{\rho}}_{\;\bar{\mu}4}. \tag{B.10}$$

Choosing $\bar{\mu} = \mu$ and $\bar{\nu} = \nu$ and multiplying the result by $g^{\mu\nu}$ yields

$$\partial_4 \Gamma^\mu_{\;\mu4} = g^{\mu\nu}\Gamma_{\bar{\rho},\nu4}\Gamma^{\bar{\rho}}_{\;\mu4}. \tag{B.11}$$

In the limit of $\Delta w \to 0$, we use (B.2), arriving at

$$-\partial_4 \Gamma^\mu_{\;\mu4} = -g^{\mu\nu}\Gamma_{4,\nu4}\Gamma^4_{\;\mu4}$$
$$= -\tfrac{1}{4}\mathcal{A}^{-1}(\partial\mathcal{A})^2, \tag{B.12}$$

which cancels out the second term in the last line (B.4), thus leaving

$$^{(5)}R^\mu_{\;4,\mu4} = -\tfrac{1}{2}\nabla_\mu(\partial^\mu\mathcal{A}), \tag{B.13}$$

or

$$g^{44\,(5)} R^\mu_{\ 4,\mu4} = -\tfrac{1}{2}\mathcal{A}^{-1} \nabla_\mu(\partial^\mu \mathcal{A}). \tag{B.14}$$

At this stage, we may put $\mathcal{A} = B^2$ in taking the limit $\Delta w \to 0$ in (1.64).

The same technique cannot be applied directly to the first term in the last line on the right-hand side of (B.5). However, we can take advantage of the symmetry properties of the curvature tensor, expecting

$$g^{\mu\nu\,(5)} R^4_{\ \mu,4\nu} = g^{44\,(5)} R^\mu_{\ 4,\mu4}, \tag{B.15}$$

and hence

$$\mathcal{L}_4 = \tfrac{1}{2}\sqrt{-g}\,B^{(5)} R = \sqrt{-g}\left(\tfrac{1}{2}BR + Bg^{44} R^\mu_{\ 4,\mu4}\right)$$
$$= \sqrt{-g}\left[\tfrac{1}{2}BR - \tfrac{1}{2}B^{-1}\nabla_\mu(\partial^\mu \mathcal{A})\right]. \tag{B.16}$$

The second term on the right-hand side is put, after integrating by parts, into the form

$$-\sqrt{-g}\,B^{-1}\nabla_\mu(B\,\partial^\mu B) = \sqrt{-g}\left(\partial_\mu B^{-1}\right)(B\,\partial^\mu B)$$
$$= -\sqrt{-g}\,B^{-1}(\partial B)^2, \tag{B.17}$$

indicating that the field $B(x)$ is a normal (nonghost) field corresponding to the choice $\epsilon = +1$. This is contrasted with the absence of this term in the five-dimensional KK theory, though $\epsilon = +1$ does not seem to be supported by a cosmological application developed in Chapter 4.

Equation (B.17) can be put into the canonical form of a kinetic term by choosing

$$B = \tfrac{1}{8}\phi^2. \tag{B.18}$$

Using this in the first term on the right-hand side of (B.16), we find

$$\xi^{-1} = 8, \quad \text{or} \quad \omega = 2. \tag{B.19}$$

Appendix C
The field equation of gravity in the presence of nonminimal coupling

We derive the field equation which replaces Einstein's equation in the presence of nonminimal coupling. Consider the Lagrangian

$$\mathcal{L}_{\mathrm{nm}} = \sqrt{-g}\,\varphi R, \tag{C.1}$$

where $\varphi(s)$ is a scalar function. Just as we do with the standard Einstein equation, we split R into R_1 that contains second derivatives of the metric and incorporate the rest into R_2:

$$R = R_1 + R_2, \tag{C.2}$$

where

$$R_1 = g^{\mu\nu}\left(\partial_\lambda \Gamma^\lambda{}_{\mu\nu} - \partial_\nu \Gamma_\mu\right), \tag{C.3}$$

$$R_2 = \Gamma_\lambda C^\lambda + \tfrac{1}{2}\Gamma^\lambda{}_{\rho\sigma}\,\partial_\lambda g^{\rho\sigma}. \tag{C.4}$$

We defined

$$\Gamma_\mu = \Gamma^\rho{}_{\rho\mu} = \frac{\partial_\mu \sqrt{-g}}{\sqrt{-g}} = \frac{1}{2}g^{\rho\sigma}\,\partial_\mu g_{\rho\sigma}, \tag{C.5}$$

$$C^\lambda = g^{\mu\nu}\Gamma^\lambda{}_{\mu\nu} = -\partial_\sigma g^{\lambda\sigma} - g^{\lambda\nu}\Gamma_\nu. \tag{C.6}$$

In deriving the last term of (C.4) we first note that R_2 contains

$$g^{\mu\nu}\Gamma^\rho{}_{\lambda\nu}\Gamma^\lambda{}_{\rho\mu} = g^{\mu\nu}\Gamma^\rho{}_{\lambda\nu}g^{\lambda\sigma}\Gamma_{\sigma,\rho\mu}. \tag{C.7}$$

Owing to the fact that $\Gamma^\rho{}_{\lambda\nu} = \Gamma^\rho{}_{\nu\lambda}$, we find that $g^{\mu\nu}\Gamma^\rho{}_{\lambda\nu}g^{\lambda\sigma}$ in (C.7) is symmetric with respect to μ and σ. For this reason only the $\mu\sigma$-symmetric part of $\Gamma_{\sigma,\rho\mu} \equiv g_{\sigma\lambda}\Gamma^\lambda{}_{\rho\mu}$ in (C.7), namely $\tfrac{1}{2}\partial_\rho g_{\sigma\mu}$, contributes to (C.4).

We will now show that

$$\sqrt{-g}R_1 = \partial_\lambda\left(\sqrt{-g}G^\lambda\right) - 2\sqrt{-g}R_2 \qquad \text{(C.8)}$$

holds, where

$$G^\lambda = C^\lambda - g^{\lambda\mu}\Gamma_\mu. \qquad \text{(C.9)}$$

This can be verified by first integrating (C.3) by parts;

$$\sqrt{-g}R_1 = \partial_\lambda\left[\sqrt{-g}\left(C^\lambda - g^{\mu\lambda}\Gamma_\mu\right)\right]$$
$$-\Gamma^\lambda{}_{\mu\nu}\,\partial_\lambda(\sqrt{-g}g^{\mu\nu}) + \Gamma_\mu\,\partial_\nu(\sqrt{-g}g^{\mu\nu}). \qquad \text{(C.10)}$$

In the first term in the second line of this equation we use (C.5), obtaining

$$\partial_\lambda(\sqrt{-g}g^{\mu\nu}) = (\partial_\lambda\sqrt{-g})g^{\mu\nu} + \sqrt{-g}\,\partial_\lambda g^{\mu\nu}$$
$$= \sqrt{-g}\,(\Gamma_\lambda g^{\mu\nu} + \partial_\lambda g^{\mu\nu}), \qquad \text{(C.11)}$$

from which it follows that

$$-\Gamma^\lambda{}_{\mu\nu}\,\partial_\lambda(\sqrt{-g}g^{\mu\nu}) = \sqrt{-g}\left(-\Gamma_\lambda C^\lambda - \Gamma^\lambda{}_{\mu\nu}\,\partial_\lambda g^{\mu\nu}\right). \qquad \text{(C.12)}$$

We also use (C.6) in the second term of the second line of (C.10), yielding

$$\partial_\nu(\sqrt{-g}g^{\mu\nu}) = (\partial_\nu\sqrt{-g})g^{\mu\nu} + \sqrt{-g}\,\partial_\nu g^{\mu\nu}$$
$$= \sqrt{-g}(\Gamma_\nu g^{\mu\nu} + \partial_\nu g^{\mu\nu}) = -\sqrt{-g}C^\mu. \qquad \text{(C.13)}$$

Using (C.12) and (C.13) in (C.11), and rearranging terms, we obtain (C.8).

Substituting (C.8) into (C.1) yields

$$\mathcal{L}_{\mathrm{nm}} = \varphi\left[\partial_\lambda\left(\sqrt{-g}G^\lambda\right) - \sqrt{-g}R_2\right]. \qquad \text{(C.14)}$$

By integrating by parts we obtain

$$\mathcal{L}_{\mathrm{nm}} = -\sqrt{-g}\left(G^\lambda\,\partial_\lambda\varphi + \varphi R_2\right). \qquad \text{(C.15)}$$

We are now going to vary this with respect to $g_{\rho\sigma}$. To the portion R_2, we apply the standard technique as explained in many textbooks:

$$\frac{\delta(\sqrt{-g}R_2)}{\delta g_{\rho\sigma}} = \sqrt{-g}G^{\rho\sigma}. \qquad \text{(C.16)}$$

We then move to the first term of (C.15):

$$\frac{\delta\left(\sqrt{-g}G^\lambda\,\partial_\lambda\varphi\right)}{\delta g_{\rho\sigma}} = \partial_\lambda\varphi\,\frac{\partial\left(\sqrt{-g}G^\lambda\right)}{\partial g_{\rho\sigma}} - \partial_\mu\left(\partial_\lambda\varphi\,\frac{\partial\left(\sqrt{-g}G^\lambda\right)}{\partial\partial_\mu g_{\rho\sigma}}\right). \qquad \text{(C.17)}$$

The computation can be simplified if we go to a local Lorentz frame. Since G^λ depends linearly on first derivatives of the metric as indicated in (C.9), we find

$$\frac{\partial\left(\sqrt{-g}G^\lambda\right)}{\partial g_{\rho\sigma}} = 0. \tag{C.18}$$

We also obtain

$$-\partial_\mu\left(\partial_\lambda\varphi\,\frac{\partial\left(\sqrt{-g}G^\lambda\right)}{\partial\partial_\mu g_{\rho\sigma}}\right) = -\sqrt{-g}\,\partial_\mu\left(\partial_\lambda\varphi\,\frac{\partial G^\lambda}{\partial\partial_\mu g_{\rho\sigma}}\right). \tag{C.19}$$

From (C.6) and (C.5) we find

$$\frac{\partial C^\lambda}{\partial\partial_\mu g_{\rho\sigma}} = g^{\rho\lambda}g^{\sigma\mu} - \frac{1}{2}g^{\rho\sigma}g^{\mu\lambda}, \tag{C.20}$$

where symmetry between ρ and σ is understood. Likewise we obtain

$$\frac{\partial}{\partial\partial_\mu g_{\rho\sigma}}\left(-g^{\lambda\nu}\Gamma_\nu\right) = -\frac{1}{2}g^{\lambda\mu}g^{\rho\sigma}. \tag{C.21}$$

By using these results we arrive at

$$\frac{\partial G^\lambda}{\partial\partial_\mu g_{\rho\sigma}} = g^{\rho\lambda}g^{\sigma\mu} - g^{\rho\sigma}g^{\mu\lambda}. \tag{C.22}$$

In this way (C.17) can be put into the form

$$\frac{\delta\left(\sqrt{-g}G^\lambda\,\partial_\lambda\varphi\right)}{\delta g_{\rho\sigma}} = -\sqrt{-g}\,\partial_\mu\left[\partial_\lambda\varphi\left(g^{\rho\lambda}g^{\sigma\mu} - g^{\rho\sigma}g^{\mu\lambda}\right)\right]$$
$$= -\sqrt{-g}(\partial^\rho\partial^\sigma - g^{\rho\sigma}\,\square)\varphi, \tag{C.23}$$

where $\square = g^{\mu\nu}\partial_\mu\partial_\nu$ is a D'Lambertian operator in Minkowski space-time.
By extending this result to curved space-time, (C.23) becomes

$$-\sqrt{-g}(\nabla^\rho\nabla^\sigma - g^{\rho\sigma}\,\square)\varphi, \tag{C.24}$$

where $\square$ is defined by

$$\square = g^{\mu\nu}\,\nabla_\mu\nabla_\nu, \tag{C.25}$$

which can be shown to agree with (2.14) when it is applied to a scalar field. (C.24) is -2 times the last term on the right-hand side of (2.6) in the text.

Appendix D
The law of conservation of matter

This appendix shows a detailed derivation of the covariant conservation law of matter. For later convenience we also show how the law changes if there is a direct coupling to matter of the scalar field in the Lagrangian. The coupling term, if there is one, is represented, for the moment, as another term,

$$L' = -J\phi, \tag{D.1}$$

where J is the source of the scalar field. Corresponding to this term, the right-hand side of Einstein's equation acquires an additional term,

$$T'_{\mu\nu} = -g_{\mu\nu}J\phi, \tag{D.2}$$

which also implies an additional term in the trace of the energy–momentum tensor;

$$T' = -4J\phi. \tag{D.3}$$

The field equation (2.13) for the scalar field is then modified to

$$\xi\phi R + \epsilon \Box\phi - J = 0. \tag{D.4}$$

Multiplying this by ϕ, we find that (2.15) is also modified to

$$2\varphi R + \epsilon\phi \Box\phi - \phi J = 0. \tag{D.5}$$

Furthermore, (2.17) is replaced by

$$-2\varphi R = T + T' - \epsilon (\partial\phi)^2 - 6 \Box\varphi. \tag{D.6}$$

By eliminating terms of R from these equations we obtain

$$\Box\varphi = \zeta^2(T - 3\phi J), \tag{D.7}$$

which replaces (2.20), where ζ^2 remains the same as before, given by (2.21) or (2.22). Equation (2.23) is also modified;

$$\nabla_\mu T^{\mu\nu} = -\nabla_\mu T_\phi^{\mu\nu} - \nabla_\mu T'^{\mu\nu} + 2(G^{\mu\nu}\,\partial_\mu + [\nabla^\nu, \Box])\varphi. \tag{D.8}$$

Consider the first term on the right-hand side. By using (2.12) we obtain

$$\nabla_\mu T_\phi^{\mu\nu} = \epsilon\left[\Box\phi\,\nabla^\nu\phi + \nabla^\mu\phi\,(\nabla_\mu\nabla^\nu\phi) - \tfrac{1}{2}\nabla^\nu(\nabla^\mu\phi\,\nabla_\mu\phi)\right]. \tag{D.9}$$

On applying (D.4), the first two terms become

$$(J - \xi\phi R)\,\nabla^\nu\phi + \epsilon(\nabla^\mu\nabla^\nu\phi)\,\partial_\mu\phi. \tag{D.10}$$

Notice that ϵ has disappeared from the first term. Also the third term on the right-hand side of (D.9) is put into the form

$$-\epsilon(\nabla^\nu\nabla^\mu\phi)\,\partial_\mu\phi. \tag{D.11}$$

We find that (D.11) differs from the second term in (D.12) only by the order in which ∇^μ and ∇^ν appear. Obviously this does not matter because we have a scalar field as the operand; hence a cancelation occurs, leaving

$$\nabla_\mu T_\phi^{\mu\nu} = J\,\nabla^\nu\phi - R\,\nabla^\nu\varphi, \tag{D.12}$$

where $\varphi = \tfrac{1}{2}\xi\phi^2$ has been used, as shown by (1.9).

As for the second term in (D.8), we have

$$\nabla_\mu T'^{\mu\nu} = -\nabla^\nu(J\phi). \tag{D.13}$$

We finally consider the third term on the right-hand side of (D.8). We compute

$$\begin{aligned}
[\nabla^\nu, \Box]\varphi &= g^{\rho\sigma}(\nabla^\nu\nabla_\rho\nabla_\sigma - \nabla_\rho\nabla^\nu\nabla_\sigma)\varphi \\
&= g^{\rho\sigma}[\nabla^\nu, \nabla_\rho]\,\nabla_\sigma\varphi \\
&= -g^{\rho\sigma}R^\lambda{}_{\sigma,}{}^\nu{}_\rho\,\partial_\lambda\varphi = -R^{\lambda\nu}\,\partial_\lambda\varphi.
\end{aligned} \tag{D.14}$$

On going from the first line to the second line we used the fact that ∇^ν and ∇_σ are commutable when they are applied to φ. We also used

$$[\nabla_\mu, \nabla_\nu]V_\kappa = -R^\lambda{}_{\kappa,\mu\nu}V_\lambda,$$

for a vector field V_κ.

By substituting (D.12), (D.13), and (D.14) into (D.8), we find that terms containing $G_{\mu\nu}$ and R cancel each other out, leaving

$$\nabla_\mu T^{\mu\nu} = \phi\,\nabla^\nu J. \tag{D.15}$$

Only for $J = 0$ do we obtain the conservation law

$$\nabla_\mu T^{\mu\nu} = 0, \tag{D.16}$$

which agrees with (2.8).

Appendix E

Eddington's parameters

Eddington's parameters, which represent the deviation from the pure version of Einstein's theory, are defined by re-expressing the Schwarzschild solution in the *isotropic coordinate system*. We start by giving the three-dimensional portion of the conventional Schwarzschild metric in the original representation:

$$ds_3^2 = \frac{dr^2}{1 - a_g/r} + r^2 \, d\Omega^2. \tag{E.1}$$

We try to determine the radial coordinate r' in terms of which (E.1) is put into the explicitly isotropic representation

$$ds_3^2 = f^2(r)\left(dr'^2 + r'^2 \, d\Omega^2\right) = f^2(r) \, d\vec{x}'^2. \tag{E.2}$$

This will be achieved if we find a function $f(r)$ such that

$$\frac{dr}{\sqrt{1 - a_g/r}} = f(r) \, dr' \tag{E.3}$$

and

$$r = f(r)r'. \tag{E.4}$$

Differentiate (E.4) with respect to r', obtaining

$$\frac{dr}{dr'} = f' \frac{dr}{dr'} r' + f.$$

Substituting (E.3) for dr/dr' and re-arranging terms yields

$$r\frac{f'}{f} = 1 - \frac{1}{\sqrt{1 - a_g/r}},$$

or

$$(\ln f)' = \frac{1}{r} - \frac{1}{\sqrt{r^2 - a_g r}}. \tag{E.5}$$

On integrating this we obtain

$$f = \frac{2r}{r - a_g/2 + \sqrt{r^2 - a_g r}}. \tag{E.6}$$

The factor 2 in the numerator was originally an integration constant, but was determined in such a way as to satisfy the condition $r' \rightarrow r$ in the limit $r \rightarrow \infty$. In fact we find $f \rightarrow 1$, a condition that is in accordance with (E.4).

On combining (E.6) and (E.4) we obtain

$$r' = \frac{1}{2}\left(r - \frac{a_g}{2} + \sqrt{r^2 - a_g r}\right), \tag{E.7}$$

which is inverted to give

$$r = r'\left(1 + \frac{a_g}{4r'}\right)^2. \tag{E.8}$$

Then (E.4) shows also that

$$f = \left(1 + \frac{a_g}{4r'}\right)^2. \tag{E.9}$$

By using this in (E.2) we obtain

$$ds_3^2 = \left(1 + \frac{a_g}{4r'}\right)^4 d\vec{x'}^2. \tag{E.10}$$

We now move on to g_{00} expressed in the form

$$-g_{00} = 1 - \frac{a_g}{r} = 1 - \frac{a_g}{r'}\frac{1}{[1 + a_g/(4r')]^2},$$

from which it follows that

$$-g_{00} = \frac{[1 - a_g/(4r')]^2}{[1 + a_g/(4r')]^2}. \tag{E.11}$$

It is also convenient to put (E.10) into the form

$$g_{ij} = \delta_{ij}\left(1 + \frac{a_g}{4r'}\right)^4. \tag{E.12}$$

We have so far stayed within general relativity. We now try to express the effect of deviation from it by means of expansion with respect to a_g/r'. First expand

$$\left(1 + \frac{a_g}{4r'}\right)^4 \approx 1 + \frac{a_g}{r'} + \mathcal{O}\left(\frac{a_g^2}{r'^2}\right) \tag{E.13}$$

and replace it by

$$\rightarrow 1 + \gamma\frac{a_g}{r'} + \mathcal{O}\left(\frac{a_g^2}{r'^2}\right). \tag{E.14}$$

Leaving only up to the first-order term, we write

$$g_{ij} = \left(1 + \gamma\frac{a_g}{r'}\right)\delta_{ij}, \tag{E.15}$$

which goes back to general relativity in the limit of $\gamma = 1$.

We apply the same technique to g_{00}:

$$-g_{00} \approx 1 - \frac{a_g}{r'} + \frac{1}{2}\frac{a_g^2}{r'^2}, \tag{E.16}$$

with the replacement

$$\rightarrow 1 - \frac{a_g}{r'} + \frac{\beta}{2}\frac{a_g^2}{r'^2}. \tag{E.17}$$

It may appear that another coefficient α should be introduced into the second term in the last line. This modification, however, can always be absorbed into the redefined a_g, thus leaving the coefficient in question to be unity as it stands.

We now conclude that any spherically symmetric solution is expressed by (E.17) and (E.15) up to the second-order approximation with respect to a_g^2/r^2 in the isotropic coordinate system. Notice, however, that we may drop terms of a_g^2/r^2 in g_{ij}, because they will affect still-higher-order terms alone when we are analyzing solar-system experiments.

In practice, however, it might often prove convenient to work with the original coordinate system. For this reason let us transform (E.15) and (E.17) back to the expression in the original coordinate system.

We first observe that (E.15) can be obtained simply by making the replacement $a_g \rightarrow \gamma a_g$ in the result from general relativity. We therefore apply the same replacement to (E.8), followed by an expansion with respect to a_g/r', obtaining

$$r = r'\left(1 + \gamma\frac{a_g}{4r'}\right)^2 \approx r' + \frac{\gamma}{2}a_g, \tag{E.18}$$

from which it follows that

$$r' \approx r - \frac{\gamma}{2} a_g = r\left(1 - \frac{\gamma}{2}\frac{a_g}{r}\right). \tag{E.19}$$

Substituting this into (E.1) results in

$$g_{rr} \approx 1 + \gamma\frac{a_g}{r}, \tag{E.20}$$

where the second-order term has been dropped, as before. The same substitution into (E.17) yields

$$-g_{00} \approx 1 - \frac{a_g}{r}\left(1 + \frac{\gamma}{2}\frac{a_g}{r}\right) + \frac{\beta}{2}\frac{a_g^2}{r^2}$$

$$= 1 - \frac{a_g}{r} + \frac{\beta - \gamma}{2}\frac{a_g^2}{r^2}. \tag{E.21}$$

Equations (E.20) and (E.21) are the desired results.

Appendix F

Conformal transformation of a spinor field

The preliminary discussion of incorporating spinor fields into curved space-time was presented briefly in section 2.5 of Chapter 2. We show in this appendix first that the theory of a massless spinor field is conformally invariant if there is no *torsion*, a concept that is central, together with curvature, to the Riemann–Cartan geometry.

The kinetic term of (2.101) is reproduced by

$$\mathcal{L}_K = -b\overline{\psi}\left(b^{\mu i}\gamma_i D_\mu\right)\psi \equiv -b\overline{\psi}\, \slashed{D}\psi, \tag{F.1}$$

where

$$b = \det\left(b_\mu^i\right) = \sqrt{-g}, \tag{F.2}$$

$$\gamma_{ij} = -\gamma_{ji} = \tfrac{1}{2}[\gamma_i, \gamma_j], \tag{F.3}$$

while the covariant derivative D_μ is now given correctly, ignoring the coupling to gauge fields for the moment, by

$$D_\mu = \partial_\mu + \tfrac{1}{4}\omega^{ij}{}_{,\mu}\gamma_{ij}. \tag{F.4}$$

In the assumed absence of torsion, the *spin connection* $\omega^{ij}{}_{,\mu}$ is given by

$$\omega_{ij,\mu} = b_\mu^k \omega_{ij,k}, \tag{F.5}$$

$$\omega_{ij,k} = -\omega_{ji,k} = \tfrac{1}{2}(\Delta_{k,ij} - \Delta_{i,jk} + \Delta_{j,ik}), \tag{F.6}$$

where *Ricci's rotation coefficient* $\Delta_{k,ij}$ is defined by

$$\Delta_{k,ij} = -\Delta_{k,ji} = \left(b_i^\mu b_j^\nu - b_j^\mu b_i^\nu\right)\partial_\nu b_{k\mu}. \tag{F.7}$$

It is convenient to re-write (F.1) into an antisymmetrized form by integrating by parts:

$$\mathcal{L}_K = -\tfrac{1}{2}b\overline{\psi}\left(\slashed{D} - \overleftarrow{\slashed{D}}\right)\psi, \tag{F.8}$$

where we used the definition

$$\overleftarrow{D} = \left(\overleftarrow{\partial}_\mu - \tfrac{1}{4}\omega^{ij}{}_{,\mu}\gamma_{ij}\right)b^{\mu i}\gamma^i. \tag{F.9}$$

Now we compare (2.95) with (3.1), finding that the corresponding conformal transformation for the tetrad is given by

$$b^i_{*\mu} = \Omega b^i_\mu. \tag{F.10}$$

We in fact verify by substituting this into (2.95) that the result reproduces the transformation rule (3.1) for the metric. Likewise we find

$$b^{\mu i}_* = \Omega^{-1} b^{\mu i}. \tag{F.11}$$

In (F.1) we have $b^{\mu i}$, for which we substitute

$$b^{\mu i} = \Omega b^{\mu i}_*, \tag{F.12}$$

as obtained from (F.11). We then have

$$bb^{\mu i} = \Omega^{-3} b_* b^{\mu i}_*. \tag{F.13}$$

In order to remove the factor Ω^{-3} we introduce the redefined spinor fields ψ_* and $\overline{\psi}_*$ defined by

$$\Omega^{-3/2}\psi = \psi_*, \qquad \Omega^{-3/2}\overline{\psi} = \overline{\psi}_*. \tag{F.14}$$

The terms of $\partial_\mu \ln \Omega = -\zeta \, \partial_\mu \sigma$ now cancel each other out in the antisymmetric combination in (F.8), providing an advantage over the complications ensuing for the scalar field. We are left, however, with those coming from $(\tfrac{1}{4})\omega^{ij}{}_{,\mu}\gamma_{ij}$.

We first derive from (F.10)

$$b^i_\mu = \Omega^{-1} b^i_{*\mu}, \tag{F.15}$$

which is substituted into (F.7) and then into (F.6), giving

$$\omega_{ij,k} = \Omega[\omega_{*ij,k} + (\eta_{ik}\,\partial_{*j} - \eta_{jk}\,\partial_{*i})\ln\Omega], \tag{F.16}$$

where we used the notation

$$\partial_{*i} = b^\mu_{*i}\,\partial_\mu. \tag{F.17}$$

Ω that occurs with the first term on the right-hand side of (F.16) disappears when we form $\omega_{ij,\mu}$ in (F.4), hence proving the invariance. Other terms coming from the second term on the right-hand side of (F.16) are found to be

$$\left\{\gamma^{ij}, \gamma^k\right\}\left(\eta_{ik}\,\partial_{*j} - \eta_{jk}\,\partial_{*i}\right)\ln\Omega = 2\left\{\gamma^{ij}, \gamma_i\right\}\partial_{*j}\ln\Omega. \tag{F.18}$$

It follows from (F.3) that

$$\{\gamma_{ij}, \gamma_k\} = 2\gamma_{ijk}. \tag{F.19}$$

The right-hand side is a completely antisymmetric product of three γ's, thus resulting in

$$\gamma^{ij}{}_i = 0. \tag{F.20}$$

For this reason (F.18) vanishes automatically. In this way (F.8) has been shown to be invariant under (F.10) and (F.14).

Next consider the mass term, (2.101):

$$\mathcal{L}_{\text{mass}} = -bm\overline{\psi}\psi. \tag{F.21}$$

By using (F.14), we immediately obtain

$$\mathcal{L}_{\text{mass}} = -\Omega^{-1}b_*m\overline{\psi}_*\psi_*. \tag{F.22}$$

The non-invariance implied by the occurrence of Ω^{-1} is due to the lack of $b^{\mu i}$, which was present in the invariant Lagrangian (F.1). Equation (F.22) may also be written as

$$\mathcal{L}_{\text{mass}} = -b_*m_*\overline{\psi}_*\psi_*, \tag{F.23}$$

in terms of a σ-dependent mass

$$m_* = m\Omega^{-1} = me^{-\zeta\sigma}. \tag{F.24}$$

Notice that this agrees with (3.63), which was obtained by choosing a scalar field for the matter field.

Combining (F.8) and (F.21) we have

$$\mathcal{L}_{\text{matter}} = -b\overline{\psi}\left[\left(\not{D} - \overleftarrow{\not{D}}\right) + m\right]\psi, \tag{F.25}$$

which is re-expressed in the E frame as

$$\mathcal{L}_{\text{matter}} = -b_*\overline{\psi}_*\left[\left(\not{D}_* - \overleftarrow{\not{D}}_*\right) + me^{-\zeta\sigma}\right]\psi_*. \tag{F.26}$$

The field equations can be derived:

$$\left(\not{D}_* + me^{-\zeta\sigma}\right)\psi_* = 0, \tag{F.27}$$

$$\overline{\psi}_*\left(-\overleftarrow{\not{D}}_* + me^{-\zeta\sigma}\right) = 0. \tag{F.28}$$

The σ equation can be derived much more easily than for the scalar field, in the sense that $\mathcal{L}_{\text{matter}}$ contains no derivative of σ. We obtain

$$\Box_*\sigma = -\zeta m_*\overline{\psi}_*\psi_*. \tag{F.29}$$

Some complication is unavoidable, however, if we try to express the right-hand side in terms of the trace of the energy–momentum tensor of the field ψ.

Corresponding to the usual definition of $T_{\mu\nu}$

$$\frac{\partial\mathcal{L}_{\text{matter}}}{\partial g^{\mu\nu}} = -\frac{1}{2}\sqrt{-g}T_{\mu\nu}, \tag{F.30}$$

we introduce $T^k{}_\mu$ defined by

$$\frac{\partial\mathcal{L}_{\text{matter}}}{\partial b^k{}_\mu} = -bT^k{}_\mu. \tag{F.31}$$

We may derive

$$T_{\nu\mu} = b_{k\nu}T^k{}_\mu, \tag{F.32}$$

which is not necessarily symmetric unless no torsion is present.

Using (F.26) we derive

$$T^k{}_\mu = \tfrac{1}{2}\overline{\psi}_*\left(\gamma^k D_{*\mu} - \overleftarrow{\slashed{D}}_{*\mu}\gamma^k\right)\psi_* \tag{F.33}$$

and

$$T_{\nu\mu} = \tfrac{1}{2}\overline{\psi}_*\left(\gamma_\nu D_{*\mu} - \overleftarrow{\slashed{D}}_{*\mu}\gamma_\nu\right)\psi_*, \tag{F.34}$$

where

$$\gamma_\nu = b_{\nu k}\gamma^k. \tag{F.35}$$

We finally obtain

$$\begin{aligned} T_* &= \tfrac{1}{2}\overline{\psi}_*\left(\slashed{D}_* - \overleftarrow{\slashed{D}}_*\right)\psi_* \\ &= -m_*\overline{\psi}_*\psi_*, \end{aligned} \tag{F.36}$$

where we used (F.27) and (F.28) in the last step. Equation (F.29) is then put into the form

$$\Box_*\sigma = \zeta T_*, \tag{F.37}$$

which agrees with (3.67), as promised.

In (F.26) one may expand $e^{-\zeta\sigma}$ with respect to σ. The first term is simply a usual mass term, while the next linear term gives the Yukawa coupling term;

$$L_{\psi\sigma} = m\overline{\psi}_* \psi_* \zeta \sigma. \tag{F.38}$$

In (F.26) we have terms of infinitely higher orders in σ. Notice that they are higher-order also in $\sqrt{G}$, as is evident from (2.22).

Since this is an entirely "usual" interaction, we can easily obtain a potential arising from exchanging a scalar particle σ between two fermions;

$$V_\sigma = -m_1 m_2 \zeta^2 \frac{1}{4\pi r}, \tag{F.39}$$

which agrees with (2.53).

We present here a spinor-field analog of the relation (4.129):

$$\mathcal{L}_{\mathrm{mass}} = -b f_\psi \phi \overline{\psi}\psi. \tag{F.40}$$

From (F.22) we find

$$\mathcal{L}_{\mathrm{mass}} = -\Omega^{-1} f_\psi b_* \phi \overline{\psi}_* \psi_*,$$
$$= -\xi^{-1/2} f_\psi b_* \overline{\psi}_* \psi_*, \tag{F.41}$$

from which can readily be given the mass term

$$\mathcal{L}_{\mathrm{mass}} = -b_* m_\psi \overline{\psi}_* \psi_*, \tag{F.42}$$

where

$$m_\psi = \xi^{-1/2} f_\psi. \tag{F.43}$$

Appendix G
Conformal transformation of the curvature scalar

We derive the rule by which the curvature scalar R in D dimensions is conformally transformed. The calculation is basically simple, but tedious and lengthy. There is, however, a convenient method. After having read this appendix quickly, the reader is advised to take a careful look at the last part. In this appendix, we express the indices in higher dimensions without using the overbars, because no confusion is expected to occur.

The Christoffel symbol is given by (3.9):

$$\Gamma^{\mu}{}_{\nu\lambda} = \tfrac{1}{2} g^{\mu\rho}(\partial_{\nu} g_{\rho\lambda} + \partial_{\lambda} g_{\rho\nu} - \partial_{\rho} g_{\nu\lambda}). \tag{G.1}$$

From this we derive R by putting

$$R = g^{\mu\nu}\left(\partial_{\lambda}\Gamma^{\lambda}{}_{\mu\nu} - \partial_{\nu}\Gamma_{\mu} - \Gamma^{\lambda}{}_{\sigma\nu}\Gamma^{\sigma}{}_{\mu\lambda}\right) + \Gamma_{\sigma}C^{\sigma}, \tag{G.2}$$

the same as (C.2)–(C.4), where we also quote (C.5) and (C.6):

$$\Gamma_{\mu} = \Gamma^{\rho}{}_{\mu\rho}, \tag{G.3}$$
$$C^{\sigma} = g^{\mu\nu}\Gamma^{\sigma}{}_{\mu\nu}. \tag{G.4}$$

Consider

$$g_{\mu\nu} = \Omega^{-2} g_{*\mu\nu}, \qquad g^{\mu\nu} = \Omega^{2} g_{*}^{\mu\nu}, \tag{G.5}$$

for which we obtain

$$\partial_{\mu} g_{\nu\lambda} = \Omega^{-2}(\partial_{\mu} g_{*\nu\lambda} - 2f_{\mu} g_{*\nu\lambda}), \tag{G.6}$$

as we already showed in (3.13), where

$$f_{\mu} = \frac{\partial_{\mu}\Omega}{\Omega} = \partial_{\mu} \ln \Omega \equiv \partial_{\mu} f. \tag{G.7}$$

209

We substitute (G.6) into (G.1), obtaining

$$\Gamma^{\mu}{}_{\nu\lambda} = \tfrac{1}{2}g_*^{\mu\rho}(\partial_\nu g_{*\rho\lambda} - 2f_\nu g_{*\rho\lambda} + \partial_\lambda g_{*\rho\nu} - 2f_\lambda g_{*\rho\nu} - \partial_\rho g_{*\nu\lambda} + 2f_\rho g_{*\nu\lambda})$$
$$= \Gamma^{\mu}{}_{*\nu\lambda} - g_*^{\mu\rho}(f_\nu g_{*\rho\lambda} + f_\lambda g_{*\rho\nu} - f_\rho g_{*\nu\lambda})$$
$$= \Gamma^{\mu}{}_{*\nu\lambda} - \delta^{\mu}_{\lambda}f_\nu - \delta^{\mu}_{\nu}f_\lambda + f^{\mu}_* g_{*\nu\lambda}. \tag{G.8}$$

This is (3.10).

It then follows that

$$\Gamma_\mu = \Gamma_{*\mu} - Df_\mu, \tag{G.9}$$

where dimensionality D has emerged through the relation

$$g_*^{\mu\nu}g_{*\mu\nu} = \delta^{\mu}_{\mu} = D. \tag{G.10}$$

In the same way we obtain

$$C^\sigma = \Omega^2[C^\sigma_* + (D-2)f^\sigma_*]. \tag{G.11}$$

On differentiating (G.8), we find

$$\partial_\lambda \Gamma^\lambda{}_{\mu\nu} = \partial_\lambda \Gamma^\lambda{}_{*\mu\nu} - \partial_\nu f_\mu - \partial_\mu f_\nu + g_{*\mu\nu}\left(\partial_\lambda f^\lambda_*\right) + (\partial_\lambda g_{*\mu\nu})f^\lambda_*. \tag{G.12}$$

Multiply this by $g^{\mu\nu}$. We thereby use

$$g_*^{\mu\nu}(\partial_\lambda g_{*\mu\nu}) = 2\Gamma_{*\lambda}, \tag{G.13}$$

which is obtained from (C.5). We thus obtain

$$g^{\mu\nu}\partial_\lambda \Gamma^\lambda{}_{\mu\nu} = \Omega^2\left[g_*^{\mu\nu}\left(\partial_\lambda \Gamma^\lambda{}_{*\mu\nu}\right) - 2g_*^{\mu\nu}(\partial_\mu f_\nu) + D(\partial_\mu f^\mu_*) + 2\Gamma_{*\lambda}f^\lambda_*\right]. \tag{G.14}$$

From (G.9) we obtain likewise

$$g^{\mu\nu}\partial_\nu \Gamma_\mu = \Omega^2(g_*^{\mu\nu}\partial_\nu \Gamma_{*\mu} - Dg_*^{\mu\nu}\partial_\nu f_\mu). \tag{G.15}$$

We move on to the terms in (G.2) containing no derivatives:

$$-g^{\mu\nu}\Gamma^\lambda{}_{\sigma\nu}\Gamma^\sigma{}_{\mu\lambda} = -\Omega^2\left[g_*^{\mu\nu}\Gamma^\lambda{}_{*\sigma\nu}\Gamma^\sigma{}_{*\mu\lambda} - 2C^\lambda_* f_\lambda - (D-2)f^\lambda_* f_\lambda\right], \tag{G.16}$$

$$\Gamma_\mu C^\mu = \Omega^2\left[\Gamma_{*\mu}C^\mu_* + (D-2)\Gamma_{*\mu}f^\mu_* - DC^\mu_* f_\mu - D(D-2)f_\mu f^\mu_*\right]. \tag{G.17}$$

By combining the terms above we finally obtain

$$R = \Omega^2 R_* + \Omega^2\Big[D\,\partial_\mu f^\mu_* + (D-2)g_*^{\mu\nu}\partial_\nu f_\mu$$
$$+ D\Gamma_{*\mu}f^\mu_* - (D-2)C^\mu_* f_\mu - (D-1)(D-2)f_\mu f^\mu_*\Big]. \tag{G.18}$$

Let us use the symbol ΔR for the terms other than the first term on the right-hand side of (G.18).

We first focus on $\partial_\nu f_\mu$, which we put into the form

$$\partial_\nu f_\mu = \partial_\nu \left(g_{*\mu\lambda} f_*^\lambda \right) = g_{*\mu\lambda} \partial_\nu f_*^\lambda + (\partial_\nu g_{*\mu\lambda}) f_*^\lambda, \qquad (G.19)$$

thence obtaining

$$g_*^{\mu\nu} \partial_\nu f_\mu = \partial_\lambda f_*^\lambda + (C_*^\mu g_{*\mu\lambda} + \Gamma_{*\lambda}) f_*^\lambda, \qquad (G.20)$$

where we have used

$$g_*^{\mu\nu} (\partial_\nu g_{*\mu\lambda}) = -g_{*\mu\lambda}(\partial_\nu g_*^{\nu\mu})$$
$$= g_{*\mu\lambda}(C_*^\mu + g_*^{\mu\nu}\Gamma_{*\nu}) = g_{*\mu\lambda}C_*^\mu + \Gamma_{*\lambda}. \qquad (G.21)$$

We also used (C.6) in the course of calculating (G.21).

Thanks to (G.20) we thus obtain

$$\Delta R = \Omega^2[2(D-1)(\partial_\mu + \Gamma_{*\mu})f_*^\mu - (D-1)(D-2)f_\mu f_*^\mu]. \qquad (G.22)$$

Now we write

$$\Box_* f = \frac{1}{\sqrt{-g_*}}(\sqrt{-g_*} f^\mu) = \partial_\mu f_*^\mu + \frac{\partial_\mu \sqrt{-g_*}}{\sqrt{-g_*}} f_*^\mu, \qquad (G.23)$$

which can be further transformed by using (C.5):

$$\Box_* f = (\partial_\mu + \Gamma_{*\mu})f_*^\mu. \qquad (G.24)$$

By choosing $f = \ln \Omega$, and using (G.24) in (G.22), we finally obtain

$$\Delta R = \Omega^2[2(D-1)\Box_* f - (D-1)(D-2)g_*^{\mu\nu} f_\mu f_\nu], \qquad (G.25)$$

which agrees with (3.14) for $D = 4$.

We add that the calculation can be simplified if we do it in the local Lorentz frame in the E frame. This implies that we drop derivatives of $g_{*\mu\nu}$ and drop $\Gamma_{*\mu\nu}^\lambda$, though its derivatives are retained. By doing so the last term on the right-hand side of (G.14) can be made to drop out. Equation (G.16) remains as it is, but only the $g_*^{\mu\nu} f_\mu f_\nu$ terms in (G.16) and (G.17) survive. As a result, terms involving $\Gamma_{*\mu}C_*^\mu$ disappear from ΔR in (G.17), simplifying the computation considerably. Likewise, the second term in (G.19) as well as $\Gamma_{*\mu}$ in (G.22) are absent. The same is true also for (G.24). In this way the same result (G.25) can be derived much more easily. The validity of this procedure is assumed owing to the tensor nature of the equation.

As a convenient continuation, we will sketch briefly how the process of moving to the E frame, as described by (3.16)–(3.34), will be modified in D dimensions. By combining (G.18) and (G.25), we find that (3.19) is now replaced by

$$\mathcal{L}_1 = \tfrac{1}{2}\sqrt{-g_*}F(\phi)\Omega^{2-D}[R_* + 2(D-1)\,\square_* f - (D-1)(D-2)g_*^{\mu\nu}f_\mu f_\nu],$$
(G.26)

where we used

$$\sqrt{-g} = \Omega^{-D}\sqrt{-g_*}$$
(G.27)

in place of (3.8), which resulted in the exponent $2 - D$ of Ω in (G.26) instead of -2. We also suppressed the overall coefficient $\mathcal{C}$ in front of $\mathcal{L}_1$ as shown explicitly in Appendix A. The condition for the absence of the nonminimal coupling term is then

$$F(\phi)\Omega^{2-D} = 1,$$
(G.28)

or

$$\Omega = F^{1/(D-2)},$$
(G.29)

which replaces (3.21). Consequently, (3.22) is replaced by

$$f_\mu = \frac{1}{D-2}\frac{F'}{F}\,\partial_\mu\phi.$$
(G.30)

We then find that the coefficient $\tfrac{3}{2}$ in (3.23) is replaced by $(D-1)/(D-2)$.

On the other hand, the second line on the right-hand side of (3.24) remains the same, on using (G.28). We then obtain the same term as (3.25) but with (3.26) replaced by

$$\Delta = \frac{D-1}{D-2}\left(\frac{F'}{F}\right)^2 + \epsilon\frac{1}{F}.$$
(G.31)

The new field σ is introduced in the same way as in (3.28) to arrive at the same result as (3.29).

For the simplest choice (3.17), we obtain the same result as the last term of (3.30) but with ζ^{-2} replaced by

$$\zeta_D^{-2} = 4\frac{D-1}{D-2} + \epsilon\xi^{-1}.$$
(G.32)

We correspondingly replace (3.39) by

$$\phi = \xi^{-1/2} e^{\zeta_D \sigma}.$$

(G.33)

Also, using (G.29) yields

$$\Omega = \exp\left(\frac{2}{D-2}\zeta_D \sigma\right),$$

(G.34)

which replaces (3.40).

Appendix H

A special choice for conformal invariance

In Chapter 3, we studied a special choice $\epsilon = -1, \xi = \frac{1}{6}$ resulting in $\Delta = 0$. In the E frame we have (3.45), giving no degree of freedom for the scalar field.

The J-frame Lagrangian (2.1) with the matter part removed now takes the form

$$\mathcal{L}_{\text{BDc}} = \sqrt{-g}\left(\frac{1}{12}\phi^2 R + \frac{1}{2}g^{\mu\nu}\,\partial_\mu\phi\,\partial_\nu\phi\right). \tag{H.1}$$

In Chapter 1 we showed that this part of the Lagrangian with ξ and ϵ arbitrary is invariant under *global* transformations (1.73) and (1.74). We find, however, that (H.1) is invariant under *local* versions, as will be shown.

Consider the transformations

$$g_{\mu\nu} = \Omega^{-2}g'_{\mu\nu}, \tag{H.2}$$

$$\phi = \Omega\phi', \tag{H.3}$$

which are essentially (1.73) and (1.74), but now with a local function Ω. We do not impose a relation like (3.20).

From (H.3) it follows that

$$\partial_\mu\phi = \Omega(\partial_\mu\phi' + \phi'\,\partial_\mu f), \tag{H.4}$$

where

$$f = \ln\Omega. \tag{H.5}$$

On substituting these into (H.1), we find

$$\mathcal{L}_{\text{BDc}} = \sqrt{-g'}\left(\frac{1}{12}\phi'^2\left(R' + \Box'f - 6g'^{\mu\nu}f_\mu f_\nu\right)\right.$$

$$\left. + \frac{1}{2}\left(\partial_\mu\phi' + \phi'\,\partial_\mu f\right)\left(\partial_\nu\phi' + \phi'\,\partial_\nu f\right)\right). \tag{H.6}$$

214

In calculating each term, we obtain, for example,

$$\tfrac{1}{2}\sqrt{-g'}\phi'^2\,\Box' f = -\tfrac{1}{2}\sqrt{-g'}\phi'\,\partial_\mu\phi'\,g'^{\mu\nu}\left(\phi^{-1}\,\partial_\nu\phi - \phi'^{-1}\,\partial_\nu\phi'\right), \qquad \text{(H.7)}$$

where we have used partial integration. We also find

$$f_\mu f_\nu = \phi^{-2}\,\partial_\mu\phi\,\partial_\nu\phi - 2\phi^{-1}\phi'^{-1}\,\partial_\mu\phi\,\partial_\nu\phi' + \phi'^{-2}\,\partial_\mu\phi'\,\partial_\nu\phi'. \qquad \text{(H.8)}$$

On substituting all of them into (H.6), we find that terms of the form $\partial_\mu\phi\,\partial_\nu\phi$ and $\partial_\mu\phi\,\partial_\nu\phi'$ cancel each other out, leaving finally the kinetic term of ϕ'. In this way we arrive at

$$\mathcal{L}_{\text{BDc}} = \sqrt{-g'}\left(\frac{1}{12}\phi'^2 R' + \frac{1}{2}g'^{\mu\nu}\,\partial_\mu\phi'\,\partial_\nu\phi'\right), \qquad \text{(H.9)}$$

which turns out to be the same as what we obtain by attaching the prime everywhere in (H.1). In this sense the theory is invariant for (H.2) and (H.3), an example of a conformal invariance, though it is not clear how this is useful in any practical application. The scalar field does not have any dynamical degree of freedom, like a gauge function in the theory of a vector field.

Appendix J

The matter energy–momentum nonconservation law in the E frame

It might be helpful if we offer a simplified derivation of (4.83). Let us ignore the curvature of space-time, and consider time derivatives only. Let the total energy, the σ energy, and the matter energy be denoted by $\mathcal{E}$, $E = \rho_\sigma$, and W, respectively. We have

$$E = \tfrac{1}{2}\dot{\sigma}^2 + V. \qquad (\text{J.1})$$

The σ field equation (4.81) may then be written as

$$-\ddot{\sigma} - V' = \zeta T_* \equiv \mathcal{F}. \qquad (\text{J.2})$$

The Bianchi identity applied to the left-hand side of (4.80) assures the covariant conservation of $T_{\mu\nu}$ on the right-hand side, which translates into

$$\dot{\mathcal{E}} = 0 = \dot{E} + \dot{W} \qquad (\text{J.3})$$

in the simplified situation.

For the first term on the right-hand side we have, according to (J.1),

$$\dot{E} = \dot{\sigma}(\ddot{\sigma} + V'), \qquad (\text{J.4})$$

the right-hand side of which is, on account of (J.2),

$$-\dot{\sigma}\mathcal{F}, \qquad (\text{J.5})$$

while the left-hand side of (J.4) is $-\dot{W}$ due to (J.3). Covariantize $\dot{W}$ and the right-hand side (J.5) yields (4.83).

Appendix K
A modification to the Λ term

We try to see how the J-frame solution considered in subsection 4.4.2 can be modified by replacing the Λ term in (4.60) by a potential

$$U_q = \Lambda \phi^q, \tag{K.1}$$

where q need be neither an integer nor positive, as suggested in (4.125). As we alluded before, it is even simpler to obtain the E-frame solutions first, and then bring them back to the J frame by a conformal transformation.

The conformal transformation moving to the E frame is the same as before, specifically with

$$\Omega = \xi^{1/2}\phi = e^{\zeta\sigma}, \tag{K.2}$$

as given by (4.95) and (4.96). In this way we arrive at the same E-frame Lagrangian (4.79) but with the resulting potential term

$$V_q(\sigma) = \Lambda \xi^{-q/2} e^{-(4-q)\sigma}, \tag{K.3}$$

which can be obtained from the potential in (4.79) by the replacement

$$\zeta \to \zeta' = \left(1 - \frac{q}{4}\right)\zeta. \tag{K.4}$$

As long as we require that σ tends to infinity asymptotically, we must limit ourselves to

$$q < 4. \tag{K.5}$$

We find attractor solutions with the same behavior as (4.99):

$$a_*(t_*) = t_*^{1/2}, \tag{K.6}$$

217

while (4.106)–(4.108) remain true under the replacement (K.4). We give specifically

$$\sigma(t_*) = \bar{\sigma} + \frac{2}{4-q}\zeta^{-1}\ln t_*. \tag{K.7}$$

The constraint (4.112) is then changed to

$$\zeta > \frac{1}{2 - q/2}. \tag{K.8}$$

As one of the easiest ways to go back to the J frame, we use (4.88) and (4.89):

$$dt_* = \Omega\, dt, \tag{K.9}$$

$$a_* = \Omega a. \tag{K.10}$$

We substitute (K.7) into (K.2) and then further into (K.9) and (K.10). After straightforward calculations, we find

$$t = A_q t_*^{(2-q)/(4-q)}, \tag{K.11}$$

$$a(t) = t^p, \quad \text{with} \quad p = \frac{1}{2}\frac{q}{q-2}, \tag{K.12}$$

if $q \neq 2$, but

$$t = A_2 \ln t_*, \tag{K.13}$$

$$a(t) = \exp\left(-\Lambda^2\xi^{-1/2}\zeta t\right), \tag{K.14}$$

if $q = 2$. The coefficient A_q is a constant given in terms of Λ, ξ, and q.

We are particularly interested in the solution (K.12). The exponent p is larger than 1 for $2 < q < 4$, whereas it is negative for $0 < q < 2$. For this reason we focus on the range $q < 0$. The exponent, however, never reaches $\frac{1}{2}$ for any finite value of q; to obtain 0.45, for example, requires $q = -18$. This indicates that it is still hard to understand the success of the standard cosmology even approximately in terms of the J-frame solutions expected from a simple potential of the type (K.1).

Appendix L
Einstein's equation in the brane world

Gravity in a brane world is somewhat different from that in KK theories, although both models are constructed in higher-dimensional space-time. In both models, Newtonian gravity must be recovered as an effective four-dimensional theory. Nevertheless, there are several differences because extra dimensions could be much larger than the Planck scale in the brane-world scenario. In particular, the five-dimensional Randall–Sundrum model [32, 33] contains a warped factor (1.45), which does not provide a simple dimensional reduction, resulting in intrinsic differences from Einstein's gravity. In the following we show these differences explicitly, assuming the Randall–Sundrum type-II model because the model is simple and concrete [89].

Since the massless graviton is confined on the brane $\mathcal{B}$, it can be described by the *intrinsic* (or *induced*) metric of the brane. We denote the unit vector normal to $\mathcal{B}$ by $n^{\bar{\mu}}$ and define the induced metric on $\mathcal{B}$ by $g_{\bar{\mu}\bar{\nu}} = {}^{(5)}g_{\bar{\mu}\bar{\nu}} - n_{\bar{\mu}}n_{\bar{\nu}}$. The barred indices are for five-dimensional space-time, as in Appendix A.

Gravity in five-dimensional space-time is described just by the five-dimensional version of Einstein's equation in the vacuum with a cosmological constant ${}^{(5)}\Lambda$ (< 0):

$$ {}^{(5)}R_{\bar{\mu}\bar{\nu}} - \tfrac{1}{2} {}^{(5)}g_{\bar{\mu}\bar{\nu}} \, {}^{(5)}R + {}^{(5)}\Lambda \, {}^{(5)}g_{\bar{\mu}\bar{\nu}} = 0. \qquad (\text{L.1}) $$

In order to derive the equation for the intrinsic metric $g_{\mu\nu}$, we decompose (L.1) into those in four-dimensional brane space-time and in its normal. The *extrinsic* curvature $K_{\mu\nu}$, which is defined by [42]

$$ K_{\mu\nu} = g_{\mu}{}^{\bar{\rho}} g_{\nu}{}^{\bar{\sigma}} \, \nabla_{\bar{\rho}} n_{\bar{\sigma}}, \qquad (\text{L.2}) $$

describing how the four-dimensional brane is embedded in five-dimensional space-time, plays a key role in this decomposition (see Fig. L.1). Note that

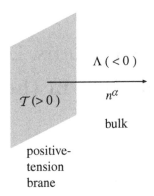

Fig. L.1. The Randall–Sundrum type-II model. Since gravity is confined on the brane $\mathcal{B}$, it can be described by the induced metric on $\mathcal{B}$.

$g_\mu{}^{\bar{p}} \neq \delta_\mu{}^{\bar{p}}$, for example, for the induced metric, unlike in the original metric ${}^{(5)}g_\mu{}^{\bar{p}}$.

Reduction of the five-dimensional curvature tensor into four-dimensional variables can be performed also by using the Gauss equation:

$$R^\mu{}_{\nu,\rho\sigma} = {}^{(5)}R^{\bar{\alpha}}{}_{\bar{\beta},\bar{\gamma}\bar{\delta}}g_{\bar{\alpha}}{}^\mu g_\nu{}^{\bar{\beta}} g_\rho{}^{\bar{\gamma}} g_{\bar{\sigma}}{}^{\bar{\delta}} + K^\mu{}_\rho K_{\nu\sigma} - K^\mu{}_\sigma K_{\nu\rho}, \qquad \text{(L.3)}$$

and the Codacci equation:

$$D_\nu K_\mu{}^\nu - D_\mu K = {}^{(5)}R_{\bar{\alpha}\bar{\beta}}n^{\bar{\alpha}}g_\mu{}^{\bar{\beta}}, \qquad \text{(L.4)}$$

where $K = K^\mu_\mu$ is the trace of extrinsic curvature, while D_μ is the covariant derivative with respect to the induced metric $g_{\mu\nu}$ rather than the original metric ${}^{(5)}g_{\bar{\mu}\bar{\nu}}$ which defines $\nabla_{\bar{\mu}}$.

We contract the Gauss equation (L.3) and use the five-dimensional version of Einstein's equation (L.1), together with the decomposition of the Riemann tensor into Weyl curvature, the Ricci tensor, and the curvature scalar:

$$ {}^{(5)}R_{\bar{\mu}\bar{\alpha},\bar{\nu}\bar{\beta}} = \tfrac{2}{3}\left(g_{\bar{\mu}[\bar{\nu}}{}^{(5)}R_{\bar{\beta}]\bar{\alpha}} - g_{\bar{\alpha}[\bar{\nu}}{}^{(5)}R_{\bar{\beta}]\bar{\mu}}\right) - \tfrac{1}{6}g_{\bar{\mu}[\bar{\nu}}g_{\bar{\beta}]\bar{\alpha}}{}^{(5)}R + {}^{(5)}C_{\bar{\mu}\bar{\alpha}\bar{\nu}\bar{\beta}}, \qquad \text{(L.5)}$$

where subscript [ ] implies an anti-symmetrization. In this way we obtain the effective four-dimensional equation:

$$G_{\mu\nu} = -\tfrac{1}{2}g_{\mu\nu}{}^{(5)}\Lambda + KK_{\mu\nu} - K_\mu{}^\sigma K_{\nu\sigma} - \tfrac{1}{2}g_{\mu\nu}\left(K^2 - K^{\alpha\beta}K_{\alpha\beta}\right) - E_{\mu\nu}, \qquad \text{(L.6)}$$

where

$$E_{\mu\nu} \equiv {}^{(5)}C_{\bar{\alpha}\bar{\mu}\bar{\beta}\bar{\nu}}n^{\bar{\alpha}}n^{\bar{\beta}}g_\mu{}^{\bar{\mu}}g_\nu{}^{\bar{\nu}}. \qquad \text{(L.7)}$$

Note that $E_{\mu\nu}$ is traceless.

For convenience, we choose a Gaussian normal coordinate y such that the hypersurface $y = 0$ coincides with the brane world, thus putting the five-dimensional metric near the brane into the simplified form:

$$ds^2 = dy^2 + g_{\mu\nu}\, dx^\mu\, dx^\nu. \qquad (L.8)$$

The singular confinement of the matter field on the brane leads us to what is called Israel's junction condition [125],

$$[g_{\mu\nu}] = 0\,, \qquad [K_{\mu\nu}] = -\left(M_P^{(5)}\right)^{-3}\left(S_{\mu\nu} - \tfrac{1}{3}g_{\mu\nu}S\right), \qquad (L.9)$$

where $[X] \equiv \lim_{y\to+0} X - \lim_{y\to-0} X = X^+ - X^-$. The extrinsic curvature $S_{\mu\nu}$ has been defined by this discontinuity, and is shown to be related to the energy and momentum densities of matter fields on the brane:

$$S_{\mu\nu} = -\mathcal{T}g_{\mu\nu} + T_{\mu\nu}, \qquad (L.10)$$

where $\mathcal{T}$ and $T_{\mu\nu}$ are the tension and the matter energy–momentum tensor in the brane world, respectively.

The Z_2-symmetry of the brane, which we assume here, uniquely determines the extrinsic curvature of the brane in terms of the energy–momentum tensor,

$$K_{\mu\nu}^+ = -K_{\mu\nu}^- = -\tfrac{1}{2}\left(M_P^{(5)}\right)^{-3}\left(S_{\mu\nu} - \tfrac{1}{3}g_{\mu\nu}S\right). \qquad (L.11)$$

On substituting (L.11) into (L.6), we obtain the gravitational equations on the 3-brane, the space-like portion of $\mathcal{B}$ [89]:

$$G_{\mu\nu} = -\Lambda g_{\mu\nu} + T_{\mu\nu} + \left(M_P^{(5)}\right)^{-6}\pi_{\mu\nu} - E_{\mu\nu}\,, \qquad (L.12)$$

where

$$\Lambda = \tfrac{1}{2}\left({}^{(5)}\Lambda + \tfrac{1}{6}\left(M_P^{(5)}\right)^{-6}\mathcal{T}^2\right), \qquad (L.13)$$

$$G = \frac{1}{48\pi}\left(M_P^{(5)}\right)^{-6}\mathcal{T}, \qquad (L.14)$$

$$\pi_{\mu\nu} = -\frac{1}{4}T_{\mu\alpha}T_\nu{}^\alpha + \frac{1}{12}TT_{\mu\nu} + \frac{1}{8}g_{\mu\nu}T_{\alpha\beta}T^{\alpha\beta} - \frac{1}{24}g_{\mu\nu}T^2. \qquad (L.15)$$

From the Codacci equation (L.4) and the five-dimensional version of Einstein's equation (L.1), we find

$$D_\nu K_\mu{}^\nu - D_\mu K = 0, \qquad (L.16)$$

which implies the conservation law for the matter,

$$D_\nu T_\mu{}^\nu = 0. \tag{L.17}$$

Equation (L.12) is quite similar to Einstein's equation, but there are two new contributions in the effective gravitational equations; one is the quadratic term of the energy–momentum tensor, $\pi_{\mu\nu}$, and the other is some component of the five-dimensional Weyl tensor, $E_{\mu\nu}$. Although these two terms will be modified in more generic brane models, they are typical contributions in a brane-world scenario.

The contracted Bianchi identities $D^\mu G_{\mu\nu} = 0$ imply that the relation between $E_{\mu\nu}$ and $\pi_{\mu\nu}$ is

$$D^\mu E_{\mu\nu} = \left(M_{\rm P}^{(5)}\right)^{-6} D^\mu \pi_{\mu\nu}. \tag{L.18}$$

Thus $E_{\mu\nu}$ is not freely specifiable but its divergence is constrained by the matter term. Equation (L.18) is still short of fixing $E_{\mu\nu}$ completely. Rather, the effective gravitational equations on the brane are not closed but one must solve the gravitational field in the bulk at the same time, in general.

In the case of cosmology, however, assuming the Robertson–Walker metric in our brane world, we can integrate (L.18), finding the 00-component of Einstein's equation from (L.12):

$$H^2 + \frac{k}{a^2} = \Lambda + \frac{1}{3}\rho + \frac{1}{36}\left(M_{\rm P}^{(5)}\right)^{-6}\rho^2 + \frac{\mathcal{K}}{3a^4}, \tag{L.19}$$

where ρ is the total energy density of matter fields, while $\mathcal{K}$ is an integration constant, which describes "dark" radiation coming from $E_{\mu\nu}$. Effects of two new terms, i.e. the quadratic term of the energy density and "dark" radiation, on the early history of the universe have been discussed widely [88].

Appendix M
Dilatation current

Consider the Lagrangian (1.12) of the prototype BD model but with the matter part removed:

$$\mathcal{L}_{\mathrm{BDc}} = \sqrt{-g}\left(\tfrac{1}{2}\xi\phi^2 R - \tfrac{1}{2}\epsilon g^{\mu\nu}\,\partial_\mu\phi\,\partial_\nu\phi\right), \qquad (M.1)$$

which can be put into the following form by the method explained in Appendix C:

$$\mathcal{L}_{\mathrm{BDc}} = -\tfrac{1}{2}\sqrt{-g}\,\xi\left(G^\lambda\,\partial_\lambda\phi^2 + \phi^2 R_2\right) - \tfrac{1}{2}\sqrt{-g}\,\epsilon g^{\mu\nu}\,\partial_\mu\phi\,\partial_\nu\phi. \qquad (M.2)$$

To this we apply (6.26) to obtain Noether's current:

$$J^\mu = \frac{\partial \mathcal{L}_{\mathrm{BDc}}}{\partial\partial_\mu g_{\rho\sigma}}\,(-2g_{\rho\sigma}) + \frac{\partial\mathcal{L}_{\mathrm{BDc}}}{\partial\partial_\mu\phi}\,\phi. \qquad (M.3)$$

Let us begin with the term containing G^μ in $\mathcal{L}_{\mathrm{BDc}}$ in the first term on the right-hand side of (M.3). According to (C.22) we have

$$\frac{\partial\left(\sqrt{-g}\,G^\lambda\right)}{\partial\partial_\mu g_{\rho\sigma}} = \sqrt{-g}\left(g^{\rho\lambda}g^{\sigma\mu} - g^{\rho\sigma}g^{\mu\lambda}\right). \qquad (M.4)$$

By multiplying this with $g^{\rho\sigma}$ we obtain

$$g^{\rho\sigma}\frac{\partial\left(\sqrt{-g}\,G^\lambda\right)}{\partial\partial_\mu g_{\rho\sigma}} = -\sqrt{-g}\,3g^{\lambda\mu}, \qquad (M.5)$$

and hence

$$\frac{\partial\left(\sqrt{-g}\,G^\lambda\,\partial_\lambda\phi^2\right)}{\partial\partial_\mu g_{\rho\sigma}}\,g_{\rho\sigma} = -\sqrt{-g}\,3g^{\mu\lambda}\,\partial_\lambda\phi^2. \qquad (M.6)$$

223

We next move to the terms containing R_2 in $\mathcal{L}_{\mathrm{BDc}}$ in the first term on the right-hand side of (M.3). From (C.5) and (C.6) we obtain

$$\frac{\partial \Gamma_\lambda}{\partial \partial_\mu g_{\rho\sigma}} = \frac{1}{2} \delta^\mu_\lambda g^{\rho\sigma}, \qquad (\text{M.7})$$

$$\frac{\partial C^\lambda}{\partial \partial_\mu g_{\rho\sigma}} = g^{\lambda\rho} g^{\mu\sigma} - \frac{1}{2} g^{\lambda\mu} g^{\rho\sigma}, \qquad (\text{M.8})$$

which can be used when we differentiate R_2 given by (C.4). We also use

$$\frac{\partial \Gamma^\lambda{}_{\alpha\beta}}{\partial \partial_\mu g_{\rho\sigma}} = \frac{1}{2} \left(g^{\lambda\rho} \delta^\mu_\alpha \delta^\sigma_\beta + g^{\lambda\rho} \delta^\mu_\beta \delta^\sigma_\alpha - g^{\lambda\mu} \delta^\rho_\alpha \delta^\sigma_\beta \right), \qquad (\text{M.9})$$

$$\frac{\partial \partial_\lambda g^{\alpha\beta}}{\partial \partial_\mu g_{\rho\sigma}} = -g^{\alpha\rho} g^{\beta\sigma} \delta^\mu_\lambda. \qquad (\text{M.10})$$

After some calculation using (C.9) we arrive at

$$\frac{(\partial \sqrt{-g} R_2)}{\partial \partial_\mu g_{\rho\sigma}} g_{\rho\sigma} = \sqrt{-g} G^\mu. \qquad (\text{M.11})$$

On combining (M.6) and (M.11), we find

$$-2 \frac{\partial \mathcal{L}_{\mathrm{BDc}}}{\partial \partial_\mu g_{\rho\sigma}} g_{\rho\sigma} = \sqrt{-g} \xi \left(-3 g^{\mu\nu} \partial_\nu \phi^2 + G^\mu \phi^2 \right). \qquad (\text{M.12})$$

For the second term on the right-hand side of (M.3) we have

$$\frac{\partial \mathcal{L}_{\mathrm{BDc}}}{\partial \partial_\mu \phi} \phi = -\sqrt{-g} \xi G^\mu \phi^2 - \sqrt{-g} \epsilon g^{\mu\nu} \phi \, \partial_\nu \phi. \qquad (\text{M.13})$$

By adding this to (M.12) we obtain

$$J^\mu = -\frac{1}{2} \sqrt{-g} (\epsilon + 6\xi) g^{\mu\nu} \partial_\nu \phi^2. \qquad (\text{M.14})$$

Note that terms of G^μ have canceled each other out.

We now consider the matter Lagrangian

$$\sqrt{-g} \left(-\frac{1}{2} g^{\mu\nu} \partial_\mu \Phi \, \partial_\nu \Phi \right), \qquad (\text{M.15})$$

which is added to (M.2). The additional contribution will be the same as the term containing ϵ in (M.14) with $\epsilon = +1$. It then follows that we simply add the term

$$-\frac{1}{2} \sqrt{-g} g^{\mu\nu} \partial_\nu \Phi^2 \qquad (\text{M.16})$$

to (M.15).

Appendix N
Loop integrals in continuous dimensions

In D dimensions we have the one-loop integral of the form

$$I_{mn} = \int d^D k \, \frac{(k^2)^{m-2}}{(k^2 + \mathcal{M}^2)^n}.$$ (N.1)

According to power counting, this is divergent of the order $D + 2m - 4 - 2n$. Suppose we choose D such that

$$D < 2n - 2m + 4.$$ (N.2)

The integral is then convergent. The result can be analytically continued to a higher value of D.

We apply a Wick rotation:

$$k^0 = ik^D,$$ (N.3)

to move to Euclidean space of D dimensions (k^D is not k raised to the power D). Using angular variables and the magnitude k, we separate from (N.1) the volume of a unit sphere:

$$V_D = \frac{2\pi^d}{\Gamma(d)},$$ (N.4)

where $d = D/2$, and $\Gamma(d)$ is a gamma function. We are thus left with an integral over the magnitude k:

$$I_{mn} = iV_D \int_0^\infty dk \, \frac{k^{2m+2d-5}}{(k^2 + \mathcal{M}^2)^n}.$$ (N.5)

By introducing a variable t by putting

$$t = k^2/\mathcal{M}^2,$$ (N.6)

we put (N.5) into the form

$$I_{mn} = iV_D \frac{1}{2} \left(\mathcal{M}^2\right)^{m-n+d-2} \int_0^\infty dt \, \frac{t^{m+d-3}}{(1+t)^n}$$

$$= iV_D \frac{1}{2} \left(\mathcal{M}^2\right)^{m-n+d-2} B(m+d-2, n-m-d+2), \quad \text{(N.7)}$$

where an integral representation of the beta function has been used:

$$B(p,q) = \frac{\Gamma(p)\Gamma(q)}{\Gamma(p+q)} = \int_0^\infty dt \, t^{p-1}(1+t)^{-p-q}. \quad \text{(N.8)}$$

Divergences due to the integration come from poles of $\Gamma(n-m-d+2)$ at

$$d = n-m+2, n-m+3, \ldots . \quad \text{(N.9)}$$

This implies that (N.7) is convergent if D is not an integer, even though the relation (N.2) is not obeyed, thus showing that the method is different from what one expects to achieve by the naive procedure of applying a cutoff. The present method in which divergent integrals are defined in terms of the integration representation of gamma functions is based on the same concept as that used for Sato's hyper-functions [127].

In the example of (6.58) we put $m = 2, n = 1$, and $\mathcal{M} = m$.

During this calculation, all the integrals remain finite as long as d is kept off the values shown in (N.9), thus allowing us to apply "symmetric integration," such that

$$\int d^D k \, k^\mu f(k^2) = 0, \quad \text{(N.10)}$$

$$\int d^D k \, k^\mu k^\nu f(k^2) = \eta^{\mu\nu} \frac{1}{D} \int d^D k \, k^2 f(k^2). \quad \text{(N.11)}$$

These are simply the relations which express isotropy in Euclidean space, but they cannot be applied naively unless the integrals are regularized.

Appendix O

A conformal frame in which particle masses are finally constant

In the calculation in section 6.6 of Chapter 6, we ignored the time-variation of the mass of the field Φ_*, which is expected to be of secondary importance in computing $\dot{\alpha}/\alpha$. Strictly speaking, however, we should move to another conformal frame in which particle masses are constant according to our principle of selecting a physical conformal frame. We will pursue this procedure in this appendix, finding that the analysis of the text remains almost unaffected while the rate of change with time of the variable G is rather small and has still to be tested by future measurements.

We start with the basic E-frame Lagrangian (3.34) with the matter Lagrangian for a free massive complex scalar field Φ_*, for simplicity. According to (6.196), the σ-dependent mass is given by

$$m_* = m^2 e^{-2\zeta'\sigma}, \tag{O.1}$$

with a constant m, while ζ' is given by (6.197). We try to move to a new conformal frame, called the J$'$ frame:

$$g_{\mu\nu} = \Omega^{-2} g'_{\mu\nu}. \tag{O.2}$$

The Lagrangian can then be cast into the form

$$\mathcal{L}_{\mathrm{BD}} = \Omega^{-4} \sqrt{-g'} \Big[\tfrac{1}{2}\Omega^2 (R' + 6\,\Box' f - 6 g'^{\mu\nu} f_\mu f_\nu) $$
$$- \Omega^2 g'^{\mu\nu}\, \partial_\mu \bar{\Phi}_*\, \partial_\nu \Phi_* - m_*^2 \bar{\Phi}_* \Phi_* \Big], \tag{O.3}$$

where

$$f = \ln \Omega \quad \text{and} \quad f_\mu = \partial_\mu f. \tag{O.4}$$

The kinetic term of Φ_* can be made canonical by introducing

$$\Phi' = \Omega^{-1} \Phi_*. \tag{O.5}$$

227

The mass term is then

$$-m_*^2\sqrt{-g_*}\,\Omega^{-4}\bar{\Phi}_*\Phi_* = -\sqrt{-g'}\Omega^{-2}m_*^2\bar{\Phi}'\Phi'. \tag{O.6}$$

In accordance with our requirement that the mass be constant, we identify (O.6) as

$$-\sqrt{-g'}m^2\bar{\Phi}'\Phi', \tag{O.7}$$

with the constant m. This fixes Ω:

$$\Omega = e^{-\zeta'\sigma}. \tag{O.8}$$

Substituting this into (O.3), we find the term resembling the kinetic term of σ:

$$-\tfrac{1}{2}\sqrt{-g'}\left(1 - 6\zeta'^2\right)e^{2\zeta'\sigma}g'^{\mu\nu}\,\partial_\mu\sigma\,\partial_\nu\sigma, \tag{O.9}$$

which can be put into a canonical form by defining a new scalar field

$$\sigma' = \sqrt{1 - 6\zeta'^2}\,\zeta'^{-1}\left(e^{\zeta'\sigma} - 1\right), \tag{O.10}$$

where we chose the integration constant such that $\sigma = 0$ corresponds to $\sigma' = 0$.

In the relation (6.193), for example, we replace $\Delta\sigma$ by

$$\Delta\sigma' \sim \sqrt{1 - 6\zeta'^2}\,\Delta\sigma, \tag{O.11}$$

which is only slightly smaller than $\Delta\sigma$.

In (O.3) we have the term

$$\frac{1}{2}\sqrt{-g'}\Omega^{-2}R' = \frac{1}{2}\sqrt{-g'}\frac{1}{8\pi G'}, \tag{O.12}$$

where

$$8\pi G' = \Omega^2 = e^{-2\zeta'\sigma}. \tag{O.13}$$

We then find

$$\dot{G}'/G' = -2\zeta'\dot{\sigma}, \tag{O.14}$$

which is obviously very small.

On comparing this with (6.194) and (6.197) and eliminating $\Delta\sigma$, we obtain

$$\frac{\Delta G'}{G'} \approx -9\mathcal{Z}^{-1}\frac{\Delta\alpha}{\alpha}. \tag{O.15}$$

The same type of relation has been found in the literature, but from a rather different theoretical background. If we use the result $\Delta\alpha/\alpha \sim 10^{-5}$ reported in [18] together with $\Delta t \sim 10^{10}$ years and $\mathcal{Z}^{-1} \sim 1$, we find that the fractional change could be as "large" as

$$\dot{G}'/G' \sim 10^{-14}\,\text{years}^{-1}, \tag{O.16}$$

which is still below the available upper bound [49, 50] by more than an order or two of magnitude.

We should further notice that (O.16) is an average rate of variation during the whole history of the universe, and need not apply to measurement in the present epoch, as is the case in the Viking Project [49] and binary-pulsar studies [50]. For such an "instantaneous" measurement, the result should be smaller by at least a further order of magnitude, because the same is true for $\Delta\sigma$, as we find in Fig. 5.10. As a result, what we thus expect for $\dot{G}/G$ should be at most 10^{-14}–10^{-15} years^{-1}, which is too small to be detected at present.

References

[1] P. Jordan (1955). *Schwerkraft und Weltall* (Friedrich Vieweg und Sohn, Braunschweig).

[2] P. A. M. Dirac (1938). *Proc. Roy. Soc.* **A165**, 199–208.

[3] Th. Kaluza (1920). *Sitzungsber. Preuß. Akad. Wiss.* 966–972: O. Klein, (1926). *Z. Phys.* **37**, 895–906.

[4] M. Fierz (1956). *Helv. Phys. Acta,* **29**, 128–134.

[5] C. Brans and R. H. Dicke (1961). *Phys. Rev.* **124**, 925–935.

[6] R. D. Reasenberg *et al.* (1979). *Astroph. J. Lett.,* **234**, L219–L221.

[7] T. M. Eubanks *et al.* (1999). *Advances in solar system tests of gravity.* Online preprint, casa.usno.navy.mil/navnet/postscript/prd_15.ps.

[8] R. H. Dicke and H. M. Goldenberg (1967). *Phys. Rev. Lett.* **18**, 313–316.

[9] V. Wagoner (1970). *Phys. Rev.* **D1**, 3209–3216.

[10] Y. Fujii (1971). *Nature Phys. Sci.* **234**, 5–7; Y. Fujii (1972). *Ann. Phys. (N. Y.)* **69**, 494–521.

[11] J. O'Hanlon (1972). *Phys. Rev. Lett.* **29**, 137–138.

[12] R. Acharia and P. A. Hogan (1973). *Lett. Nuovo Cim.,* **6**, 668–672.

[13] T. Damour and A. M. Polyakov (1994). *Nucl. Phys.* **B423**, 532–558.

[14] A. G. Riess *et al.* (1998). *Astron. J.* **116**, 1009–1038; P. M. Garnavich *et al.* (1998). *Astrophys. J.* **509**, 74–79.

[15] S. Perlmutter *et al.* (1998). *Nature* **391**, 51–54; S. Perlmutter *et al.* (1999). *Astrophys. J.* **517**, 565–586.

[16] S. Weinberg (1989). *Rev. Mod. Phys.* **61**, 1–23.

[17] R. R. Caldwell, R. Dave and P. J. Steinhardt (1998). *Phys. Rev. Lett.* **80**, 1582–1585; L. Wang, R. R. Caldwell, J. P. Ostriker, and P. J. Steinhardt (2000). *Astrophys. J.* **530**, 17–35.

[18] J. K. Webb, M. T. Murphy, V. V. Flambaum, A. Dzuba, J. D. Barrow, C. W. Churchill, J. X. Prochaska, and A. M. Wolfe (2001). *Phys. Rev. Lett.* **87**, 091301 (4 pages).

[19] H. Weyl (1951). *Space, Time and Matter* (H. L. Brose, Tran.) (Dover, New York).

[20] Y. M. Cho and P. G. O. Freund (1975). *Phys. Rev.* **D12**, 1711–1720.

[21] J. Scherk and J. Schwarz (1979). *Nucl. Phys.* **B153**, 61–88.

[22] Y. M. Cho (1992). *Phys. Rev. Lett.* **68**, 3133–3136.

[23] M. B. Green, J. H. Schwarz, and E. Witten (1987). *Superstring Theory* (Cambridge University Press, Cambridge).

[24] C. G. Callan, D. Friedan, E. J. Martinec, and M. J. Perry (1985). *Nucl. Phys.* **262**, 593–609; C. G. Callan, I. R. Klebanov, and M. J. Perry (1986). *Nucl. Phys.* **278**, 78–90.

[25] J. D. Barrow (1996). *Mon. Not. Astron. Soc.* **282**, 1397–1406.

[26] G. T. Horowitz and A. Strominger (1991). *Nucl. Phys.* **B360**, 197–209. A membrane is first discussed as an extended object of a string in K. Kikkawa and M. Yamasaki (1986). *Prog. Theor. Phys.* **76**, 1379–1389; M. J. Duff, P. S. Howe, T. Inami, and K. S. Stelle (1987). *Phys. Lett.* **B191**, 70–74; E. Bergshoeff, E. Sezgin, and P. K. Townsend (1988). *Ann. Phys.* **185**, 330.

[27] V. A. Rubakov and M. E. Shaposhinikov (1983). *Phys. Lett.* **B125**, 139–143; K. Akama (1983). In *Gauge Theory and Gravitation*, ed. by K. Kikkawa, N. Nakanishi, and H. Nariai (Springer-Verlag, Berlin).

[28] E. Witten (1995). *Nucl. Phys.* **B443**, 85–126; see also a review by P. K. Townsend (1996), *Four lectures on M-theory*, hep-th/9612121.

[29] P. Hořava and E. Witten (1996). *Nucl. Phys.* **B460**, 506–524; (1996). *ibid* **B475**, 94–114.

[30] N. Arkani-Hamed, S. Dimopoulos, and G. Dvali (1998). *Phys. Lett.* **B429**, 263–272; I. Antoniadis, N. Arkani-Hamed, S. Dimopoulos, and G. Dvali (1998). *Phys. Lett.* **B436**, 257–263.

[31] V. P. Mitrofanov and O. I. Ponomareva (1988). *Sov. Phys. JETP* **67**, 1963–1966; C. D. Hoyle, U. Schmidt, B. R. Heckel, E. G. Adelberger, J. H. Gundlach, D. J. Kapner, and H. E. Swanson (2001). *Phys. Rev. Lett.* **86**, 1418–1421; J. C. Long, A. B. Churnside, and J. C. Price (2001). In *Proceedings of the 9th Marcel Grossmann Meeting* (World Scientific, Singapore).

[32] L. Randall and R. Sundrum (1999). *Phys. Rev. Lett.* **83**, 3370–3373.

[33] L. Randall and R. Sundrum (1999). *Phys. Rev. Lett.* **83**, 4690–4693.

[34] A. Lukas, B. A. Ovrut, K. S. Stelle, and D. Waldram (1999). *Phys. Rev.* **D59**, 086001 (9 pages); H. S. Reall (1999). *Phys. Rev.* **D59**, 103506 (6 pages); A. Lukas, B. A. Ovrut, and D. Waldram (1999). *Phys. Rev.* **D60**, 086001 (11 pages).

[35] J. Garriga and T. Tanaka (2000). *Phys. Rev. Lett.* **84**, 2778–2781.

[36] C. S. Chan, P. L. Paul, and H. Verlinde (2000). *Nucl. Phys.* **B581**, 156–164.

[37] M. J. Duff, J. T. Liu, and K. S. Stelle (2001). *J. Math. Phys.* **42**, 3027–3047.

[38] A. Kokado, G. Konishi, T. Saito, and K. Uehara (1996). *Prog. Theor. Phys.* **96**, 1291–1299; A. Kokado, G. Konishi, T. Saito, and Y. Tada (1998). *Prog. Theor. Phys.* **99**, 293–303.

[39] P. G. Roll, D. Pekár, and R. H. Dicke (1964). *Ann. Phys. (N.Y.)* **26**, 442–517.

[40] E. Fischbach and C. Talmadge (1998). *The Search for Non-Newtonian Gravity* (AIP Press, Springer-Verlag, New York).

[41] C. Will (2001). *Living Reviews in Relativity*, **4**, 4 (Albert-Einstein-Institut, Potsdam, Germany, www.livingreviews.org/Articles/).

[42] C. W. Misner, K. S. Thorn, and J. A. Wheeler (1973). *Gravitation* (Freeman, New York).

[43] Y. Fujii (1998). *Proc. International Workshop on JHF Science (JHF98)*, ed. J. Chiba, M. Furusaka, H. Niyatake, and S. Sawada (KEK, Tsukuba, Japan), pp. II 125–128; ArXiv gr-qc/9806092.

[44] R. Utiyama, private communication.

[45] K. Nordvedt, (1968). *Phys. Rev.* **169**, 1017–1015; (1995). *Icarus* **114**, 51–62.

[46] S. Weinberg (1972). *Gravitation and Cosmology* (John Wiley, New York).

[47] W.-T. Ni (1977). *Phys. Rev. Lett.* **38**, 301–304.

[48] V. Faraoni, E. Gunzig, and P. Nardone (1998). *Conformal transformations in classical gravitational theories and in coslomogy.* gr-qc/9811047.

[49] R. W. Hellings *et al.* (1983). *Phys. Rev. Lett.* **51**, 1609–1612.

[50] S. E. Thorset (1996). *Phys. Rev. Lett.* **77**, 1432–1435.

[51] L. Landau (1955). In *Niels Bohr and the Development of Physics,* ed. W. Pauli (McGraw-Hill, New York).

[52] R. H. Dicke (1959). *Science* **129**, 621–624; J. D. Bjorken (1963). *Ann. Phys. (N.Y.)* **24**, 174–187; H. Terazawa, Y. Chikashige, and K. Akama (1977). *Phys. Rev.* **D15**, 480–487.

[53] See, for example, M. E. Peshkin and D. V. Schroeder (1995). *An Introduction to Quantum Field Theory* (Addison Wesley, New York).

[54] R. Utiyama (1956). *Phys. Rev.* **101**, 1597–1607.

[55] T. W. K. Kibble (1961). *J. Math. Phys.* **2**, 212–221.

[56] K. Hayashi and A. Bregman (1973). *Ann. Phys.* **75**, 562–600.

[57] D. I. Santiago (2000). *Gen. Rel. Grav.* **32**, 565–581.

[58] Y. Fujii and T. Nishioka (1990). *Phys. Rev.* **D42**, 361–370.

[59] R. H. Dicke (1962). *Phys. Rev.* **125**, 2163–2167.

[60] S. Coleman (1988). *Nucl. Phys.* **B310**, 643–668.

[61] S. Deser and B. Zunimo (1977). *Phys. Rev. Lett.*, **38**, 1433–1436.

[62] R. D. Peccei, J. Sola, and C. Wetterich (1987). *Phys. Lett.* **B195**, 183–190.

[63] A. H. Guth (1981). *Phys. Rev.* **D23**, 347–356.

[64] K. Sato (1981). *Mon. Not. R. Astron. Soc.* **195**, 467–479; K. Sato (1981). *Phys. Lett.* **99B**, 66–70.

[65] A. D. Linde (1982). *Phys. Lett.* **108B**, 389–393.

[66] A. Albrecht and P. J. Steinhardt (1982). *Phys. Rev. Lett.* **48**, 1220–1223.

[67] A. Linde (1983). *Phys. Lett.* **B129**, 177–181.

[68] Y. Fujii and J. M. Niedra (1983). *Prog. Theor. Phys.* **70**, 412–423.

[69] Y. Fujii (1998). Preprint, arXiv:gr-qc/9609044.

[70] Y. Fujii (1998). *Prog. Theor. Phys.* **99**, 599–621.

[71] C. Wetterich (1988). *Nucl. Phys.* **B302**, 645–667.

[72] D. Dolgov (1982). In *The Very Early Universe*, Proc. Nuffield Workshop, ed. G. W. Gibbons and S. T. Siklos (Cambridge University Press, Cambridge).

[73] L. H. Ford (1987). *Phys. Rev.* **D35**, 2339–2344.

[74] J. Yokoyama and K. Maeda (1988). *Phys. Lett.* **B207**, 31–35.

[75] T. Nishioka and Y. Fujii (1992). *Phys. Rev.* **D45**, 2140–2143.

[76] M. Endō and T. Fukui (1977). *Gen. Rel. Grav.* **8**, 833–839.

[77] J. P. Ostriker and P. J. Steinhardt (1995). *Nature* **377**, 600–602.

[78] G. Eftathiou, W. J. Southerland, and S. J. Maddox (1990). *Nature* **348**, 705–707.

[79] M. Fukugita, F. Takahara, K. Yamashita, and Y. Yoshii (1990). *Astrophys. J.* **361**, L1–L4.

[80] M. Fukugita and E. L. Turner (1991). *Mon. Not. R. Astron. Soc.* **253**, 99–106.

[81] P. de Bernardis *et al.* (2000). *Nature* **404**, 955–959; A. E. Lange *et al.* (2001). *Phys. Rev.* **D63**, 042001 (8 pages).

[82] A. Balbi *et al.* (2000). *Astrophys. J.* **545**, L1–L4.

[83] K. Tomita (2001). *Prog. Theor. Phys.* **105**, 419–427; K. Tomita (2001). *Mon. Not. R. Astron. Soc.* **326**, 287–297.

[84] B. Ratra and P. J. E. Peebles (1988). *Phys. Rev.* **D37**, 3406–3427.

[85] T. Chiba, N. Sugiyama, and T. Nakamura (1997). *Mon. Not. R. Astron. Soc.* **289**, L5–L9.

[86] P. Binétruy (1999). *Phys. Rev.* **D60**, 063502 (4 pages); P. Brax and J. Martin (2000). *Phys. Rev.* **D61**, 103502 (14 pages); T. R. Taylor, G. Veneziano, and S. Yankielowicz (1983). *Nucl. Phys.* **B218**, 493–513; I. Affleck, M. Dine, and N. Seiberg (1985). *Nucl. Phys.* **B256**, 557–599.

[87] P. J. Steinhardt, L. Wang, and I. Zlatev (1999). *Phys. Rev.* **D59**, 123504 (13 pages).

[88] P. Binétruy, C. Deffayet, and D. Langlois (2000). *Nucl. Phys.* **B565**, 269–287; N. Kaloper (1999). *Phys. Rev.* **D60**, 123506 (14 pages); C. Csaki, M. Graesser, C. Kolda, and J. Terning (1999). *Phys. Lett.* **B462**, 34–40; T. Nihei (1999). *Phys. Lett.* **B465**, 81–85; P. Kanti, I. I. Kogan, K. A. Olive, and M. Prospelov (1999). *Phys. Lett.* **B468**, 31–39; J. M. Cline, C. Grojean, and G. Servant (1999). *Phys. Rev. Lett.* **83**, 4245–4248; P. Binétruy, C. Deffayet, U. Ellwanger, and D. Langlois (2000). *Phys. Lett.* **B477**, 285–291.

[89] T. Shiromizu, K. Maeda, and M. Sasaki (2000). *Phys. Rev.* **D62**, 024012 (6 pages).

[90] K. Maeda and D. Wands (2000). *Phys. Rev.* **D62**, 124009 (9 pages); C. Barceló and M. Visser (2000). *JHEP* **0010**, 019; A. Mennim and R. A. Battye (2001). *Class. Quantum Grav.* **18**, 2171–2194.

[91] I. Zlatev, L. Wang, and P. J. Steinhardt (1999). *Phys. Rev. Lett.* **82**, 896–899.

[92] A. Masiero, M. Pietroni, and F. Rosati (2000). *Phys. Rev.* **D61**, 023504 (8 pages); F. Perrotta, C. Baccigalupi, and S. Matarrese (2000). *Phys. Rev.* **D61**, 023507 (12 pages); O. Bertolami, and P. J. Martins (2000). *Phys. Rev.* **D61**, 064007 (6 pages); P. Brax and J. Martin (2000). *Phys. Rev.* **D61**, 103502 (14 pages); T. Barreiro, E. J. Copeland, and N. J. Nunes (2000). *Phys. Rev.* **D61**, 127301 (4 pages); V. Faraoni (2000). *Phys. Rev.* **D62**, 023504 (15 pages); L. Amendola (2000). *Phys. Rev.* **D62**, 043511 (10 pages); L. A. Urena-Lopez and T. Matos (2000). *Phys. Rev.* **D62**, 081302 (4 pages); V. Sahni and L. Wang (2000). *Phys. Rev.* **D62**, 103517 (4 pages); E. J. Copeland, N. J. Nunes, and F. Rosati (2000). *Phys. Rev.* **D62**, 123503 (7 pages); C. Baccigalupi, S. Matarrese, and F. Perrotta (2000), *Phys. Rev.* **D62**, 123510 (15 pages).

[93] K. Maeda (2001). *Phys. Rev.* **D64**, 123525 (4 pages); S. Mizuno and K. Maeda (2001). *Phys. Rev.* **D64**, 123521 (16 pages).

[94] S. Carroll (1998). *Phys. Rev. Lett.*, **81**, 3067–3070.

[95] P. G. Ferreira and M. Joyce (1998). *Phys. Rev.* **D58**, 023503 (23 pages).

[96] Y. Fujii and T. Nishioka (1991). *Phys. Lett.* **B254**, 347–354.

[97] Y. Fujii (1996). *Astropart. Phys.* **5**, 133–138.

[98] Y. Fujii (2000). *Phys. Rev.* **D62**, 064004 (6 pages).

[99] F. C. Adams, J. R. Bond, K. Freeze, J. A. Frieman, and A. V. Olinto (1993). *Phys. Rev.* **D47**, 426–455; W. H. Kinney, and K. T. Mahanthappa (1994). *ibid.* **52**, 5529–5537.

[100] S. V. Ketov (1997). *Fortschr. Phys.* **45**, 237–292.

[101] D. Ruelle (1989). *Chaotic Evolution and Strange Attractors* (Cambridge University Press, Cambridge); D. K. Arrowsmith and C. M. Place (1990). *An Introduction to Dynamical Systems* (Cambridge University Press, Cambridge).

[102] S. Dodelson, M. Kaplinghat, and E. Stewart (2000). *Phys. Rev. Lett.* **85** 5276–5279.

[103] S. Hellerman, N. Kalpor, and L. Susskind (2001). *JHEP* **0106**, 003; W. Fischler, A. Kashani-Poor, R. McNees, and S. Paban (2001). *JHEP* **0107**, 003; P. K. Townsend (2001). *JHEP* **0111**, 042.

[104] T. Futamase and K. Maeda (1989). *Phys. Rev.* **D39**, 393–404.

[105] R. Easther and K. Maeda (1996). *Phys. Rev.* **D54**, 7252–7260.

[106] I. Prigogine (1977). In *Nobel Lectures in Chemistry 1971–1980*, ed. S. Forsén (World Scientific Publishing Company, Singapore, 1993).

[107] Y. Fujii (1998). *XXXIIIrd Rencontre de Moriond, Fundamental Parameters in Cosmology* (Les Arcs, France, January 17–24): ArXiv gr-qc/9806089.

[108] T. Chawanya (1995). *Prog. Theor. Phys.* **94**, 163–179.

[109] H. Meinhardt (1995). *The Algorithmic Beauty of Sea Shells* (Springer-Verlag, Berlin).

[110] M. Morikawa (1990). *Astrophys. J.* **362**, L37–L39; M. Morikawa (1991). *ibid* **369**, 20–29; T. Fukuyama, M. Miyoshi, M. Hatakeyama, M. Morikawa, and A. Nakamichi (1997). *Int. J. Mod. Phys.* **D6**, 69–90.

[111] E. Fischbach and C. Talmadge (1992). *Nature* **356**, 207–215.

[112] C. Wetterich (1988). *Nucl. Phys.* **B302**, 668–696.

[113] Y. Fujii (1974). *Phys. Rev.* **D9**, 874–876; Y. Fujii (1991). *Int. J. Mod. Phys.* **6**, 3305–3357.

[114] T. Muta (1987). *Foundations of Quantum Chromodynamics* (World Scientific, Singapore).

[115] M. P. Locher (1990). *Proceedings of 12th International Conference on Particles and Nuclei (MIT)*; *Nucl. Phys.* **A527**, 73c–88c.

[116] Yasushi Takahashi, private communication.

[117] T. Chiba (1999). *Phys. Rev.* **D60**, 083508 (4 pages).

[118] N. Bartolo and M. Pietroni (2000). *Phys. Rev.* **D61**, 023518 (6 pages).

[119] Y. Fujii (2000). *Phys. Rev.* **D62**, 044011 (6 pages).

[120] E. Fischbach *et al.* (1986). *Phys. Rev. Lett.*, **56**, 3–6.

[121] R. V. Eötvös, V. Pekár, and E. Fekete (1922). *Ann. Phys. (Lpz.)* **68**, 11–66.

[122] A. I. Shlyakhter (1976). *Nature* **264**, 340–340; (1983). ATOMKI Report A/1 unpublished.

[123] T. Damour and F. J. Dyson (1996). *Nucl. Phys.* **B480**, 37–54.

[124] Y. Fujii, A. Iwamoto, T. Fukahori, T. Ohnuki, M. Nakagawa, H. Hidaka, Y. Oura and P. Möller (2000). *Nucl. Phys.* **B573**, 377–401.

[125] M. J. Drinkwater, J. K. Webb, J. D. Barrow, and V. V. Flambaum (1998). *Mon. Not. R. Astron. Soc.* **295**, 457–462; H. Terazawa (2000). KEK Preprint 99–161.

[126] W. Israel (1996). *Nuovo Cim.* **44B**, 1–14.

[127] M. Sato (1958). *Sügaku*, **10**, 1–27; M. Sato (1959–60). *J. Faculty of Sci. (University of Tokyo)*, Sec. I, **8**, 139–193; Y. Fujii and K. Mima (1977). *Prog. Theor. Phys.* **58**, 991–1006; Y. Fujii, (1978). In *Particles and Fields*, ed. D. H. Boal and A. N. Kamal (Plenum Publishing Corporation, New York).

Index